Time Before Space

An Airman's Odyssey . . . from Biplanes to Rockets

Time Before Space

An Airman's Odyssey . . . from Biplanes to Rockets

E. H. Rowley

Former Chief of Flight Test
Boeing-Wichita

Foreword by Jim Greenwood

1531 Yuma (Box 1009), Manhattan, Kansas 66502-4228 USA

Printed in the United States of America on acid-free paper.

ISBN 0-89745-174-0 paper
ISBN 0-89745-178-3 hardback

Cover: By Mike Boss

About the Artist

Michael K. Boss, of Hill City, Kansas, is a specialist in transportation art. He primarily paints detailed scenes of aircraft, railroad locomotives, ships, and space vehicles. He studied art under Jack Leynnwood of Woodland Hills, California, and attended Kansas State University and Southern Illinois University.

Boss has won various awards at both aviation and railroad showings and has the rare talent to combine the two areas. He is currently at work on a series of paintings of the Old Los Angeles Airports. Two of his paintings — *Vail Morning* and *Winnie Mae* — will appear in the 1995 calendar "Golden Age of Flight" produced by the Smithsonian Institution from the National Air & Space Museum Collection.

The cover painting of the Curtiss Oriole, the XB-47, and the refueling B-29s typifies the aviation experiences of Elton Rowley. The work was done in casein on rag illustration board.

Layout by Lori L. Daniel

To Jean and my family for their patience through the years.

Contents

Foreword ix

Acknowledgments xiii

Introduction xv

Time Before Space, by Billie Cornell xvii

Chapter

1 The Beginning 1

2 The Big Kite 5

3 Our Final Project 7

4 Sprouting Wings 9

5 The Military 13

6 Colonel John Peglo 16

7 Proving a Point 19

8 The Moment of Truth 23

9 The Big Break 25

10 Master Sergeant DePussen 31

11 The Real Poncho 36

Chapter

12	Home on Leave	44
14	The Air Mail Challenge	51
15	The *Explorer I* Project and Fox News Reel	55
16	The Great George Patton	64
17	Civilian Life in Milwaukee	69
18	Married Life	76
19	Milwaukee Airways	81
20	Curtiss-Buffalo	87
21	The Curtiss Turkey Hawk	94
22	A Difficult Decision	101
23	Spartan Aircraft	107
24	Spartan's Navy Debut	113
25	The NP-1 Project	121
26	J. Paul Getty	126
27	Boeing-Wichita	138
28	The B-29 Challenge	143
29	Post-War Boeing-Wichita	163
30	OPERATION DRIP	177
31	013 on April Fool's Day	191
32	The Flying Boom	195
33	The Jet Transition	205
34	The B-47 and Pete Jensen	211
35	The B-47 Development	217
36	Conclusion	224

Index 232

Foreword

In the course of my own aviation career, I've been associated with a number of celebrated experimental test pilots. I've always admired and respected this elite group, whose flying and technical skills — and risk-taking — continue to pave the way for better and safer airplanes. "At best," said the legendary one-time test pilot Jimmy Doolittle, "it is a hazardous business."

I began flying in 1936. Two years later, the MGM movie *Test Pilot* gave us a Hollywood look at the "business," loosely based on the writings of Jimmy Collins. He is remembered as the test pilot who grimly predicted his own death in a newspaper column published a few days after he lost his life diving a new Navy fighter.

Although I had made some live tests on new parachutes before and during World War II, not until 1950 did I get a chance to test-fly an airplane. The company where I worked had just completed a major modification to a high-time DC-3 — what they called the "Gooney Bird" in World War II. We needed a final flight test for CAA approval. Granted, it wasn't a "first flight," but it was a test hop of sorts. And I would be the copilot.

Our chief test pilot, Anson "Johnny" Johnson, a moonlighting airline captain who won the 1948 Thompson Trophy Race in a P-51 Mustang, took the left seat. Little did I know that Johnny wanted to show me the old bird's aerobatic capabilities. He really wrung it out; Donald Douglas himself would have winced. My job was to watch the gauges, but on the third split-S I lost my lunch.

Several test pilots I know have written books, good books — Tony Levier, Pete Everest, Scott Crossfield. Others might have considered writing one, but to most pilots such a task has all the appeal of an in-flight fire over the mountains at night.

Enter Elton Holcomb Rowley, a retired chief of flight test, who may not have the name recognition of others, but among his peers ranked at the top. That's a distinction not easily attained in the exclusive test pilot community. Moreover, many of his fellow pilots died in the pursuit of their profession. He is a survivor, largely due to his lifelong dedication to excellence, but also possibly due to a little luck.

Usually pilots who commit their experiences to paper have the editorial assistance of a professional writer. What sets Elton's book apart is the fact it is a *real* autobiography — he wrote every word himself.

Time Before Space is an airman's odyssey, the story of a boy who dreamed of flight and made that dream come true. It is told clearly and concisely in a kind of shirt-sleeve English all of us can comprehend. It is the warm, personal account of the author's exciting career in a period that was, perhaps, the most dynamic in aviation history.

This was the time between the "ragwings" of the 1920s and the man-carrying, rocket-powered vehicles that emerged around 1960. Those were pivotal years when the earliest experiments in flight were fading and a young, vibrant industry was beginning to mature. We saw an airport system take shape, new navigation aids evolve, and incredible gains in aircraft performance — accelerated by the exigencies of a second world war.

Elton Rowley was part of it all — in both the commercial and military spheres of activity. While his contributions to the science of flight may not have created headlines in the style of a Chuck Yeager or a Neil Armstrong, they were, nevertheless, many and significant. You cannot necessarily measure the value of a man's work by the size of his scrapbook.

Actually, the author's modesty belies his vital role in the advancement of aviation. He flew the mail, ran a flying service, and designed airplanes, including a high-performing Greve-class racer. He barnstormed, operated an airport, and, in the early '30s, was one of the most weatherwise pilots in the Army Air Corps. He also was deeply involved in upper air research.

At Curtiss-Wright he rescued the lagging SBC-4 modification program and at Spartan he saved the Navy NP-1, which later went down the tube. A nonconforming trouble-shooter who got results, his reputation for toughness also got him the nickname "Cactus." He bulldozed bureaucracies relentlessly.

Elton's persistent efforts on the B-29 and B-47 projects are credited with helping Boeing bring about engineering changes and other innovations that solved critical operational problems both aircraft suffered in their early development cycles. The B-29, of course, went on to drop the first atomic bomb, and the B-47 became the Strategic Air Command's first all-jet bomber — clearly the most advanced weapon systems platform of its day.

Indeed, the author was in the forefront of refinements that greatly improved the dissemination of weather data, instrument flying proficiency, aerial refueling techniques, flying safety criteria, and training procedures applicable to all manner of flying machines — from fabric biplanes to multiengine jets.

The author also had a talent for picking the best people for his flight-test team. He hired Alvin "Tex" Johnston, who later would achieve notoriety by barrel-rolling the 707 prototype above the heads of the crowd at the annual powerboat races in Seattle, ostensibly to demonstrate the plane's structural integrity. Although I doubt Elton would condone such a maneuver, which was witnessed by some 200,000 racing fans, I'm also confident he still rates Tex as one of the best.

Ever since he was a skinny kid jumping off hills with a homemade glider, Elton Rowley, who learned to fly in the hardest of times, sought perfection. In a challenging and demanding job all too often marked by human tragedy, he never lost his sense of purpose nor his sense of humor.

Elton had the respect of virtually everyone with whom he came in contact, from generals and industrialists to secretaries and line boys. He pioneered concepts in flight testing and in product demonstrations that are standard practice today. His outstanding flying abilities were, in fact, commended by no less a master flier than Charles A. Lindbergh, as well as a president-to-be — General Dwight D. Eisenhower.

This book is broadly informative, revealing little-known or hitherto untold facts behind important events that figured in our nation's aeronautical progress. This will not only cause some of the players to rewrite their own memoirs, it will add to the body of knowledge documenting America's aviation heritage.

The narrative also took me on my own nostalgic, sentimental journey into the past. Revisiting aviation luminaries like Earl Schaefer, Max Balfour, Ora Young, Jack Shaffer, Dick Taylor, and so many other familiar names recalled some wonderful (and some scary) memories of my early exploits in the flying game.

Elton can be justly proud of his many accomplishments. He and all pilots who boldly risk their lives by pushing aircraft beyond their limits have made the skies safer for the rest of us. While it is a well-known fact that Boeing builds great airplanes, it is the likes of the Eddie Allens, the Tex Johnstons, and the Elton Rowleys who prove their airworthiness.

So fasten your seatbelts and prepare to "ride shotgun" with Cactus as he makes the transition from the Curtiss Oriole, in which he soloed, to the progenitors of modern, computerized jet transports. It is a fascinating adventure you won't soon forget.

Jim Greenwood
Green Valley, Arizona

Acknowledgments

— To my friend N. D. Showalter who arranged for the declassification of the Boeing in-flight refueling pictures, as well as other data and news releases that are very important to the historical value of my book.

— To the National Air and Space Museum of the Smithsonian Institution for special information and pictures of early airplanes.

— To Lindsey A. Dunn, assistant director-curator of the Glenn H. Curtiss Museum of Local History, Hammondsport, New York, for his counsel, pictures, and important dates.

— To Jim Greenwood for his editing of the original manuscript along with his expert counsel on the whole book project.

— To Don McNeal of the Council Grove, Kansas, *Republican* newspaper for his assistance with the initial galley proofs and pictures.

— To Billie Cornell for the typing of the original manuscript and for the poem, "Time Before Space."

— And, finally, to those unnamed who helped in the many details to finish this project.

Introduction

After I retired from both my aviation career and the family automobile business, my family and friends insisted that I write a book about my experiences.

It was difficult for me to get started as I knew it was going to be a long, hard job. It became more and more interesting, however, when an outline and then an actual manuscript came together from my log books and pictures. It was particularly rewarding when I found and talked to old friends and associates. Some were no longer around, however, and I asked myself why I had let the years get away.

There was a time in aviation history, from Lindbergh's flight to Paris in 1927 up to the first manned space flight, when technology and aviation hardware development advanced at a rate that hasn't been equaled before or since.

World War II was a great catalyst. Its end marked the beginning of the jet and rocket age, and this major step forward gave the aviation industry the thrust to go beyond the speed of sound. As a result, the size and shape of airplanes changed: the supersonic fighters, beginning with North American's sleek F-100 Super Sabre; the high-altitude jet bombers, such as Boeing's B-47 Stratojet and B-52 Stratofortress; and the first great commercial airliner, the 707. Outer space was within our grasp.

My aviation career actually started in 1929 as a solo student and ended as the head of a large flight-test operation from 1943 to 1951, then as a project engineer from 1952 to 1958. I continued to fly privately until 1985. The period between the 1920s and 1960 is what this book is about.

Dedicated to the Men and Women who risked Life and Fortune to establish the Great Aviation Industry. They were Beacons of Light in the Embryonic Age of Flight.

THIS IS TO CERTIFY THAT

Elton H. Rowley

HAS BEEN SELECTED AS A MEMBER OF THE

OX5 AVIATION PIONEERS HALL OF FAME

AND IS AWARDED THIS HONOR IN RECOGNITION OF CONTRIBUTIONS TO THE ESTABLISHMENT OF THE AIR INDUSTRY.

By authority of Board of Governors and the undersigned officers of the OX5 AVIATION PIONEERS, THIS 22ND DAY OF SEPTEMBER, 1992

By _______________ PRESIDENT

_______________ SECRETARY

Time Before Space

He lay in the grass at the top of a hill,
dreaming the dreams that boys often will.
Watching birds soar freely in the ocean of air,
softly, so softly, he whispered a prayer.

Lord, take my hand so that someday, I
will fly like the birds in Your magical sky.
Give me knowledge and courage that I might take my place,
in what one day will be known as the "Time Before Space."

The Lord must have smiled on that boy down below,
for He lighted the way the boy wished to go.
With hand on his shoulder, He guided him through,
the dangerous tasks all test pilots must do.

He gave him a family to share in his dream,
and comfort when the challenge seemed far too extreme.
He gave the boy wisdom that he might take his place,
to become part of the legend of "Time Before Space."

Billie Cornell

The Beginning

I was born 4 August 1911 in Bristol Center, in the heart of the Finger Lakes region of upstate New York. My father, John Rowley, and my mother, Grace Holcomb, were hard-working, hardy, farm folks. My only brother, Maurice, was ten years my senior; as a result, we were not very close. My cousin, Clark Holcomb, was about my age, so we were together a lot as kids. Our home town was in a beautiful valley just over the hill from Hammondsport, where Glenn Curtiss became famous for his flying activities. Airplanes were a frequent sight, flying over the countryside and at about every public gathering or county fair, and barnstorming from fields around our town.

I think I caught the flying bug from this. I remember taking my first airplane ride in a Curtiss Jenny at about ten years old. I had the bug for sure after that experience. My cousin Clark and I read all the books and magazines we could find on airplanes. Our enthusiasm had grown to the point where we decided to build a hang glider from plans we found at our local library. But we needed help, so we talked our ideas and plans over with our local blacksmith and car repairman, Frank Tones. Frank agreed to help as long as it was okay with our folks. We assured him there would be no problem.

We had the perfect hill spotted. It was facing the prevailing wind with a flat top, a sheer drop of about 10 feet, then a gentle slope of about 500 feet with a smooth, flat landing spot at the bottom. The nice thing about this was that the hill was in my dad's pasture.

The planning was complete and work was started on the bird. My dad knew what was going on, but figured it was just boys playing and dreaming.

School vacation was about to start. In our planning, we thought it wise to do the testing when the folks were on their annual trip to the Adirondack Mountains in August. We had one month to finish the machine and to do the test-flying. But finances were a problem. We needed to buy covering material, music wire for bracing, etc. Frank loaned us the money on the strength of my allowance due before Mother and Dad would leave on their vacation.

My mother's sister, Stella, was going to look after Clark and me, as we were just 12 years old at the time. This was great because Stella knew little of our plans. My folks left on schedule and everything was ready for the big adventure. In about a week we were ready to move the bird to the hill for final assembly. We were proud of our job. The design was typical, a biplane with an opening in the lower wing for the pilot seat and hand grips on each side of the opening. The tail was mounted on the wood beams extending aft from the top and bottom wings with two vertical struts diagonally braced with music wire; a horizontal stabilizer and elevator were mounted on the top tail beam with strut bracing extending to the lower tail beam. The rudder was a fixed surface between the two tail beams. The only positive control was the elevator, actuated by a lever mounted on the right side just forward of the right hand grip.

To fly the bird, the instructions read, "Step inside wings, lift glider (about 50 lbs.), set elevator control nose slightly down. Run into the wind and jump off a small hill. Avoid winds under 15 mph or over 30. Maintain lateral balance by shifting body weight to the high side."

My concern was my feet. What if my landing was fast? I'd have to run like hell!

The instructions were clear, the wind was right, and I was ready. I backed off and ran. The tail came off before I came to the sheer of the hill. As I jumped, I thought, "This thing is nose heavy. Elevator, more nose up."

That was a mistake. The bird nosed up and slid off on one wing, and I was in a ball of sticks and fabric. I was bruised and had small cuts, though no bones were broken. But, I was a hell of a lot smarter.

The next day, Clark and I looked at the broken machine and decided to rebuild but change the control, add some more wing area, and cover the top side of the wings. The original was covered on the bottom only. The big change was the addition of wheels so that we could use a shock cord like a slingshot for launching. I could have my hands free for more control of the aileron and elevator.

The time before my folks were due back was getting short. Less than two weeks. So it was back to Frank. Could we use the loft over his main shop? We confided in him that we were going to build a new bird. "It's okay," he said, "but if your dad disagrees when he gets back, you are out for good."

We knew we had to have the bird in an advanced stage of construction before the folks returned. I figured the psychological effect would be worth it. Dad would be impressed and curious, so maybe we would have a chance to get the actual flying approved.

The folks returned from their vacation and, as predicted, Dad was interested. In a few days, all was ready to go. However, we thought it wise to wait until Mother and Dad were away again for the first flight.

The shock launch was a problem. The shock cord was not available and we were broke. Frank, as usual, came to our aid.

"There's a stack of old inner tubes out back. Why don't you cut them up in long strips and wrap the strips with cord."

This idea worked. Two days later, we had a shock cord, such as it was, a four-inch ring for it, and a hook to fasten onto the landing gear. We slid the ring over the cord to the middle and fixed it so that the ring would not slide off center, and, presto, we had all the elements of our launching system.

Not to attract too much attention, we rigged the launching gear on the hill before bringing the bird: two stakes about 40 feet apart, just back of the brink of the hill, and another stake at the limits of our shock cord stretch. We pulled the cord back to the limit and released it. *Wham!* went the ring, and to our surprise, the rebound was such that the launching ring almost came back to the release point. We were pleased. Now, how do we release the bird after it is pulled into position hooked to the shock cord?

Clark said, "Let's tie a rope around the tail skid, pull the bird back to another post, and when you're ready, I'll chop the rope with a hatchet. All you have to do is give me the word."

My brother was about to get married and school had started, so there was a delay until my folks went to Rochester for the weekend to visit my brother and his future bride's folks. Aunt Stella stayed with us again.

This was our chance to make the flight. The wind was about 10 mph. We rushed the bird to the hill. By that time, however, the word was out so we had some help. I had acquired a white cotton helmet and goggles. I have often wondered what the ten or twelve people helping and watching thought. The folks I was most concerned about were Frank Tones and Dr. McDowell. Doc had brought me into the world. I thought this just had to work, but in the back of my head I was thinking, "Doc is here in case I bust my butt!"

The bird was pulled into position. I was in the cockpit with my feet firmly on the cross-bar of the landing gear. "Goggles and helmet, check controls, don't act nervous, don't over control. Oh, hell! We'll soon know! The flight will be about one minute, at best."

I gave the word. "Release." The next thing I remember, I was 50 feet out over the hill. The bird was sinking and I thought, "More nose up. No! Nose down! Don't make the same mistake as before." With that, the bird settled into a steep glide. Landing was coming up fast. "Stall the bird! Haul back on the stick!" She settled in hard but nothing was broken. The wind was down. No more flights that day.

I think the most pleased of the town folks who witnessed the flight was Frank Tones. Rightfully so. He had made it all possible.

My father saw two or three future flights, but the last one was a disaster. The wind was gusty and slightly crossed. The takeoff was routine. The drift was toward some trees to the right of my normal flight path. Body against the drift, adverse yaw, airspeed dropping, sinking fast, trees looming up to my right. I crashed into the trees. The bird was a wreck. I wasn't injured, just minor cuts and bruises. I could see my father running down the hill with Clark, but by the time they arrived, I had recovered my composure and had freed myself from the wreckage.

My dad's first words were, "I'll be damned! The crazy machine could have killed you. I want you to understand, I'm glad you're not hurt, but this is the last of your glider flying!"

Dad's statement left me with mixed feelings. He wasn't as angry or decisive as normal, so I thought at the time that he was just a little proud of our glider capers. I'm sure Frank Tones and Dad had discussed our flying operations several times and were pleased with our work.

Clark and I decided it would be wise to abandon the glider experiments and so with my dad's knowledge, we burned the bird. I had a feeling he didn't like the dramatic gesture anymore than we did. However, this was not the end of our flying activities.

2

The Big Kite

The 1924 school year was about finished, so what was our summer program going to be? Gliders were out. Clark and I agreed that such activity was off limits, so why not build the biggest kite that had ever been constructed? We could hang a rope straight down from the center of it, and we could swing on the rope as long as the wind would blow hard enough to support either of us. This idea would take lots of rope and plenty of light muslin cloth, and a winch to reel the kite in and out. It also would require two or three people to launch it high enough to be able to use the winch. Late spring was our best wind period. We could build the kite on weekends and after school.

The kite design was the typical shape, 32 feet long and 28 feet wide. Again, money was a problem; we needed several yards of muslin 54 inches wide, 500 feet of 5/16-inch hemp rope, and main kite beams of 1¼- x 1¼-inch ash or oak. All together it would cost $60. The winch we could borrow from a well digger friend of my father.

The plan was presented to my dad and mother and they thought it was all right. Dad said, "I'll make a deal. If you and Clark will keep the corn lot cultivated this summer, I'll finance your project." We were off and running. Frank Tones agreed to make the cross-beams of ash, and the local storekeeper ordered the muslin and 500 feet of 5/16-inch high-grade hemp rope.

It was late March when the kite was ready to fly. No snow, with frequent high winds. Our field was on the flats 500 yards from Dad's barn.

All was made ready: main kite line to the winch and a guess as to the tail streamer. After many tries adjusting the tail and harness of the kite, we were successful. She was pulling hard on the winch. The vertical line down from the kite was the big problem. We had to find the center of lift and adjust the vertical line accordingly.

The people who were helping us lost interest, except for Frank Tones's nephew Lonny, who was about five years old. We had tried to hang on the vertical line, but Clark and I were too heavy. Clark said, "Lonny, why don't you try." He was all for the idea. We made a simple rope harness and fastened him to the vertical line. We reeled in the kite and off he went. He thought this was great! Clark reeled in the kite a little more and Lonny was 20 feet off the ground. Then, all hell broke loose! Lonny let out the most bloodcurdling scream; he was scared. My mother heard him, as did his mother.

We let him down and cut the rope harness to turn him loose. He ran home still screaming and crying. Mother arrived on the scene and we gave her the reason for the commotion, but this did very little good. Her stern words echoed, "Your fathers will hear of this from me. I assure you there will be appropriate punishment for you both."

That evening my parents and Lonny's father and mother and Clark's father and mother all assembled. Clark and I were on trial. We had both agreed to make light of the whole thing and make it sound like a great experience for Lonny. We assured all present that everything was perfectly safe — we had made sure of that. Things did calm down, but Clark and I got a good butt-chewing. As I expected, my father said, "How the hell did you figure that caper out? I'd like to see the big kite fly." My reply was, "Let's get everybody to watch." For the next flight we had 30 or more people. This turned into a field day, though Lonny and his folks didn't come.

For the rest of the summer, Clark and I cultivated corn and kept our noses clean. However, greater plans were taking shape.

Our Final Project

Clark and I had been working with and dreaming about airplanes since we were ten years old. We had learned a lot in four years. I can say that for my effort and with my flying experience I was ready for the real thing. As for Clark, his interests were elsewhere. We talked about what we were going to do after high school and maybe college. "For sure," he said, "I'm not going to fly. I'm going to have a business of my own, a motel-restaurant or something like that."

I said I was going to get flight training and then go into the service to get more. Clark thought we ought to dream up one more project.

We talked about the new home-built airplanes that were featured in some of the flying magazines. The Heath Parasol, light and simple, was a little bird that you could build from a kit. But money again was a problem. Where would we get $550 and a four-cylinder Henderson motorcycle engine converted for a propeller shaft?

My reaction was to design our own and build it our way. But Clark just wasn't with the program; he dropped out, so I went on alone.

I admit I copied the Heath, except my fuselage was wood-wire braced. It was the summer of 1925 when the Rowley Model 0100 was started. In the winter of 1925, the fuselage was complete, and the landing gear was installed; the fuselage had a seat, and the wheels had tires. I had scrounged a big 74-cubic-inch Harley motorcycle engine from the dealer in my high school town, Canandaigua, New York. The local Nash dealer's son, George Popwell, was in the same freshman class in mechanics. He was interested, so we worked the engine over and converted it for a propeller drive shaft and hub. I designed and built the propeller as a special project in the woodworking class.

By spring 1926, the engine was installed and the tail surfaces were completed and attached. The rudder, elevator, and aileron controls were installed and working. The engine had been installed and tested. My school grades had suffered in my freshman high school year so my father said, "No more airplanes until your grades are okay. This summer you'll go back for extra tutoring — work in the local pea vinery three days and go to school two days a week."

I was 14 so I could get my junior driver's license for school- and work-driving only. Dad gave me a used Model T Ford, which helped soften the pain of not being able to complete the airplane. Before the high school semester ended, however, my extra work was finished, so the pressure was off and the airplane came back into the picture. I couldn't resist taxiing the fuselage with the rudder and elevators working. It gave me the feel of the basic handling of a real airplane on the ground, taxiing downwind, taxiing with tail up into the wind, turning with the rudder, all using the propeller blast and forward speed control. The experience only served to strengthen my resolve to fly.

My dad insisted that my sophomore year in high school would be the start of my prep school work to prepare me for law school: math, physics, chemistry, language, and history. Mechanics and mechanical drawing were great, but the languages were a bore. Gym and track were great, too. I made the track team as a hurdler and the gymnastics team. Dramatics and public speaking were also strong subjects. The Canandaigua Academy also had a marching band. I played first trombone. Clark, my cousin, also played in the trombone section, making it real fun. My sophomore year ended on a pleasant note, and Dad and Mother thought I was finally on my way. What they didn't know was that the flying dream was there as strong as ever and plans were being made to take flight training.

The great news of the Lindbergh flight to Paris, on 20 and 21 May 1927, in my sophomore year, further strengthened my resolve to become the best pilot I could possibly be.

Sprouting Wings

I graduated from the Canandaigua Academy in June 1929 with good grades. My father still wanted me to become a lawyer. My brother told me there was a future for me in the construction business. But my thoughts were to become a pilot. Aviation was growing in spite of the serious depression our whole country was about to enter.

I had saved for my initial flight training by working summers, and after graduation began working full time for my brother.

Without discussing my real desires with the family, I enrolled in a night aviation school, the "Times Union School of Aviation," developed by Russell Holderman, famous chief corporate pilot for the Gannett newspapers of Rochester, New York. The school offered aviation math, mechanics, navigation, and weekend flight training. Ground school was three nights a week. As soon as the school schedule would allow, I started my flight training.

My choice of instructors was a World War I German Air Service pilot, Otto Enderton. He was tough, but fair. He would praise you for the good and raise hell about the mistakes. Otto had the ability to give you confidence in handling the airplane. After eight hours, I soloed a Curtiss C-6 Oriole, with ease and confidence. Otto was pleased and told me so. I decided after the ground school to continue on with Otto for more flight training, which included aerobatics and cross-country dual and solo.

By June 1931 I had about 100 hours total and Otto said, "You can make it as a pilot. My advice is to expand your training in the military." I took his advice and enlisted in July. I saw Otto many times thereafter; I made an effort to see him whenever I was in the Rochester area. Otto became the

The Curtiss Oriole. The author soloed in this type in October 1929. (Photo, The Glenn Curtiss Museum of Local History, Hammondsport, NY)

Department of Commerce Inspector (which later became the Civil Aviation Authority) and remained in that position until his retirement. He followed my career until he died.

The Operations Office and Weather Station Headquarters building at Scott Field (across street by flag pole).

The Military

I announced my decision to join the Army. My mother was the hardest to sell; it took the recruiting officer to convince her. Of course, the flying came into the picture, which didn't help. Dad conceded he wouldn't have his lawyer, and Mother finally realized that her youngest was literally flying out of the nest.

On 11 July 1931 I was sworn in at Syracuse, New York, and was transported to Fort Monmouth, New Jersey, for school screening. For what school was I most qualified? It turned out that meteorology was the best for me because after school I could be transferred to an Air Corps base. There was also a chance to become an Air Corps cadet.

After three months of boot training, the meteorology school started. We received a college degree in meteorology in nine months — school six days a week and plenty of homework. It was tough but very interesting and a real challenge. I graduated in the top 10 percent of my class and was offered an assignment at the training school as an instructor.

I elected to be transferred to Scott Field, Illinois, as a forecaster and preflight weather briefer. It was rumored that Scott was to be one of the new upper air reconnaissance flight centers. This I was interested in because of the flight activity that I would be involved with when the program started.

I arrived at Scott Field in late August 1932, having taken two weeks leave to see my folks and my girlfriend, Jean Austin. This young lady would play a great part in my future.

Scott Field was a mixed bag of lighter-than-air, heavier-than-air, and the Signal Corps detachments. I was assigned to the weather section of the Signal Corps.

The Thomas Morse 0-19, a 15th Observation Squadron aircraft used for upper-air sounding, 1933. (Photo, Smithsonian Photo Archives)

My seniority was zero, being "the new kid on the block," so I naturally got the early morning schedule, which included drawing the weather map and the early morning weather briefings. But this was good for me because I was fresh out of school and was trained in the new frontal theory of forecasting. The surface weather map that I drew reflected the new way of forecasting, and this new map received a lot of comments, both pro and con.

Shortly after my arrival, Captain Niles Schoffield, the Signal Corps Detachment Commander, called all of the weather station personnel to a meeting to announce that it was definite that Scott Field would be a weather center for upper air data. He asked if I would set up the new program when the new equipment arrived; he was talking in terms of two weeks to be operative. I gladly accepted. The other weathermen didn't like the flying part, of course, but it suited me perfectly because of my background. It meant that I would get a flight observer's rating and flight pay.

The equipment arrived — the aerometeorograph, an instrument to record temperature and pressure up to 20,000 feet, with calibration procedures and installation instructions included. Two airplanes had to be equipped. However, at the start of the program, only one was assigned. Naturally, it was the oldest with high engine time, a real "hangar queen." It was a regular squadron airplane, a Thomas Morse 0-19. These were tough birds and would go to 20,000 feet in one hour, but with open cockpits. This made cold-weather flying a real challenge.

The pilots for the new upper air program were to be from the 15th Observation Squadron; I had a different one each day. Some were good, and some were bad. Some of our data was usable, and some was marginal. All together, however, our forecasting missed about 25 percent. So I complained to Captain Schoffield that our score looked bad and it reflected on me as well as the pilots, although mechanical failure also was part of our problem.

Colonel John Peglo

While I was still a forecaster and weather briefer — a specialist 4th class and a private 1st class — assigned to early morning weather, Colonel John Peglo was also on the scene.

Colonel Peglo had been a German lighter-than-air expert during World War I under the famous Count Von Zeppelin. Like many of the young German officers, he had elected to emigrate to the United States, to continue his career in the U.S. military. Germans, as naturalized citizens, served in both the U.S. Army and the U.S. Navy.

Colonel Peglo was charged with commanding both lighter- and heavier-than-air at Scott Field, but his obvious preference was lighter-than-air. Colonel John had worked up through the ranks to his position and was the field's base commander.

My early morning assignment was a tough duty: 0400 hours, draw weather map on a large glass plate in flight operations, and give the weather briefing for Colonel John. I had not been forewarned about the Colonel, and I will never forget my first encounter.

He arrived at operations with his khaki Packard sedan with driver and his little red dachshund, "Fritzie." He stopped at the large map that I had drawn, with the front lines, both cold and warm, along with general wind-flow patterns. I was at attention and gave my name and rank. He acknowledged. I started my briefing as I had been taught. After about ten words, he turned to me and said in broken English, "Vot the hell is with this map and vot do you know about the vether?" I was as surprised at his comment as he was at the map style because I thought he had been pre-briefed about these new frontal maps in contrast to those with the usual high- and low-pressure cells. Obviously, Colonel John just wasn't accustomed to the new type of forecasting.

At the time, I thought it best to simply keep quiet and retire to the instrument room just off the main operations area. I sat down and faked working. As I left, Colonel Peglo's little dog, Fritzie, followed me. I snapped my fingers and she jumped into my lap. I was petting her, but he didn't notice until he was ready to leave. He turned, saw Fritzie, and said, "Fritzie, nein. Kommen!" She jumped down and followed him out. And that was the end of my first encounter with Colonel John.

The next day, our meeting was entirely different. There was some weather developing to the west, and it was moving fast in our direction. Colonel John arrived on schedule. This time I said, "Good morning, Sir."

His reply was, "Is it?"

"Not if you want to fly, Sir." I gave him a complete briefing; he asked questions and made remarks. At the end, I summed up our situation. "We better have all aircraft in the hangars by 1100 hours. The wind will be from the north, 35 knots, by that time. Rain and possible hail will precede the front, Sir."

He looked at me. "Good briefing. Thanks." I petted Fritzie and Colonel John left.

Staff Sergeant Finley, who was in charge of Operations, remarked, "The old man is in good humor this morning."

By 1100 hours, as predicted, we had our storm, and all aircraft were hangared. The next morning Colonel John came directly to the instrument room. He was a little early so I was still working on the briefing data. He said, "We guessed it right yesterday."

"Yes, Sir. Let's not call it guessing. You predicted it on the button, Sir," I replied.

He continued to look over the current data. I made no comment. Weather was improving. In my opinion, we should fly the next day. Colonel John completed his tour and remarked, "Improving. Tomorrow should be acceptable."

"Yes, Sir, I'll watch the situation and check with you at 1500 hours."

"Good," he replied.

For several mornings, the weather was good so Colonel John's visits were short. I noticed that Fritzie was not with him, so I finally asked about her. "She is under house arrest," he replied. "She has Fräulein troubles." With a smile he left.

Second Lieutenant Halverson had recently arrived from flying school. He had a little screw-tail Boston Bull (male). I hadn't noticed the dog back of the flight-plan counter because I was busy either taking outside instrument readings or was working in the instrument room. And Sergeant Finley apparently had been taking care of the dog for Lieutenant Halverson when he was out flying.

It was Monday morning, and Colonel John had come into Operations with Fritzie. Master Sergeant Bill Bishop was filling in for Sergeant Finley. I didn't give the little bulldog any thought. I was with Colonel John in the instrument office when I heard Sergeant Bishop say, "You little bastard, get away from Fritzie." It was too late. By the time Colonel John and I got out of the office, Fritzie and the little bulldog were in the typical canine embrace. Colonel John was really angry and started to kick the bulldog.

I yelled at him not to hurt Fritzie, so he stopped, and I then chewed out Sergeant Bishop. I got away with it because Bishop was in shock and Colonel John was there and heard my comments.

I told Colonel John that the only thing to do was wait. "When this is over, Fritzie should be taken to the nearest vet."

Colonel John said, "Hell no! Take her to the base hospital!" Sergeant Bishop volunteered, as he felt responsible for the problem.

I was glad to be off duty when Sergeant Bishop returned. However, the next morning he confronted me about the "bad remarks" I had made during the fracas the day before. He reminded me that he was the top-ranking Non Com on base and if I wanted to get along I should apologize in front of the Colonel. I agreed that I had used bad language but if he wanted to make something of it, I'd meet him in the gym. "Just take the stripes off," I said; "You're lucky Colonel John didn't bust you back to a buck."

In the long run, Fritzie didn't have puppies and everything went back to normal. From that time on, Colonel John was friendly toward me, and I felt relaxed working around him.

Proving a Point

Captain Schoffield knew that my earlier problems with obtaining quality weather traces were not improving in the Aerial Weather Recon program. He finally called me to his office to discuss the program in general. He was particularly interested in why we continued to have a bad forecasting score and asked me to outline our difficulties in order. My reply was, "First, pilot proficiency. Second, mechanical failures of the bird. Third, weather aborts because of lack of navigation and weather flying instrumentation." He promised to talk to Colonel John and would be in touch.

I figured that all hell would break loose and I would be in the middle of it, but I felt it was about time to get our problems solved, regardless of where the ball bounced.

We were in the fall of 1932. Scott Field is located on flat land about 30 miles east of the Mississippi River where ground fogs occurred frequently in fall and spring. I remember one morning we had a typical thin layer and I knew that as soon as the sun appeared, and if we got a little wind, it would disappear in minutes, so I prepared the airplane for flight and suited up. First Lieutenant Holcomb was scheduled as pilot.

Colonel John came in shortly for his early morning weather briefing. He spotted me all suited up in the pilot's lounge where the big glass weather map was located. I stood up and walked toward him. He stopped and said, "Vot the hell is wrong?"

"No pilot, Sir," I replied.

He glanced at the flight-schedule board. The fog was lifting fast, and he gave the order to make the blimps ready for flight. I waited another 30 minutes, and First Lieutenant Holcomb came into Operations. It was 0800 and too late to make our flight and get data on the teletype before the 0900 deadline.

The buck fell on Lieutenant Holcomb. He had aborted the flight on the schedule board because of the fog, yet Colonel John had been out flying the blimps later that morning. The Lieutenant had created a problem, and Sergeant Finley, Operations Non Com, commented to me, "I'd bet he will hear plenty about this from Captain Smith and Colonel John." Scott Field was CAVU (clear and visibility unlimited) that day.

The next morning, Lieutenant Ben Dalley, my tennis partner, was scheduled to pilot. The conditions were the same as the day before. I was suited up when Ben arrived. He looked a little rough and said, "Pard, it's a little sticky out there, but looks like you think it's go. You take her. I need the rest." As he climbed into the front cockpit, I replied, "Okay."

It was a perfect flight except for the blind takeoff and 30 seconds of needle, ball, and airspeed through the fog layer. But at 17,000 feet you could see for 50 miles or more! In the distance, St. Louis was a blob of smoke, as usual. After a few seconds enjoying the scenery, I pulled power, stalled the bird, and kicked left rudder. She went into a left spin after eight or ten turns. We were down to around 5,000 feet normal recovery and a steep spiral to traffic altitude. Then landing. When we were taxiing to the line, I heard Ben say, "You're stealing my stuff." I replied, "It's the best and fastest way down, so why not?"

We had the aerometeorograph in 45 minutes early, and I suggested we look at the trace. Ben studied it for a few seconds and said, "You're cheating. You can't fly that good." He slapped me on the back. "Let's play a little tennis after lunch." I asked if he could handle the strain and he replied, "Oh, hell, that cold air brought me back to full speed."

After lunch I walked across the lawn to Ben's quarters. I found him ready to go. It seemed I couldn't miss on this day; I skunked him four games straight. He threw his racquet up in the air and said, "Rowley, you do everything like you're killing snakes."

"Ben," I replied, "let's apply the same approach to our weather recon program."

Ben agreed, "Pard, you're probably right. I know what you were trying to do this morning. You wanted to prove a point. I'll admit we all could use some practice on needle, ball, and airspeed. But you did a no-no on the blind takeoff and then the fog layer. Some of the pilots won't push it that far."

"Well, Ben, if we keep the abort rate the way it is, you can bet we're going to have some serious visitors and Colonel John won't help much because you know what he thinks about heavier-than-air operations." Colonel John still didn't like heavier-than-air one bit. He felt the heavier-than-air people were not as careful with their flight planning — even sloppy.

I continued, "One thing I know, the better our weather data, the more accurate our forecasting will be. This thing is a national challenge, and we'd better get serious."

Ben replied, "Got your point, Pard."

Second Lieutenant Dutch Kleinoder, 15th Observation Squadron weather pilot, Scott Army Air Base, 1932-1933.

8

The Moment of Truth

Later that day, Captain Schoffield sent his aide — what we called his "top kick" — to tell me to be in Colonel John's office the following day at 0900 — my day off. To my surprise, Major Koenig, Colonel John's vice commander; Captain Smith, the 15th Observation Squadron commander; and Captain Schoffield already were there. I figured, here it comes! They all obviously had been talking about my complaints. I asked the Sergeant Major outside the door what was cooking. His reply: "Better knock and go in."

I went through the typical greeting procedure — attention, salute, serial number; then Colonel John said, "At ease. Take a seat."

Colonel John opened the meeting. "Tell us about the Upper Air Recon program." I figured this was it. As briefly as possible, I went through what could be done to improve our overall efficiency, as I had done with Captain Schoffield. Colonel John had my file in front of him. As I finished, he said, "You have had pilot training. Be honest with me. How much of the actual flying have you been doing?"

"All, except with Lieutenants Brownfield, Holcomb, Kleinoder, and Shedd," I replied.

"Who does the best job of this group?"

"Lieutenant Kleinoder, Sir," I replied.

Colonel John looked at me with a twinkle in his eye and said, "Thanks for a good briefing. You are dismissed."

The Monday after Colonel John's briefing, Lieutenant Dutch Kleinoder was assigned the flight. Weather was good, so everything went off without a hitch. As we were walking from the line afterwards, Dutch said, "Thanks for the good word Saturday."

The lighter-than-air hangar at Scott Army Air Base, IL, looking west, 1933. (Photo, U.S. Signal Corps Photo Section, Section, Scott Field, IL)

I noted, "Word gets around here fast."

Dutch said, "If you play your cards right, you can have the whole program. If that's what you're bucking for, Ben Dalley and I agree. You have what it takes, but be ready for the flak and extra work that goes with it."

I questioned him. "Dutch, what was Captain Smith's reaction to the briefing?"

"Hell, Rowley, the 15th Squadron needs this weather flight duty like a hole in the head. Smitty's reaction was positive; he feels you're doing him a favor." Dutch added, "Mark my word, you'll hear more soon."

The Big Break

As Lieutenant Kleinoder had predicted, that day after the weather flight, Sergeant Finley in Operations told me to meet Captain Schoffield in his office at 1000 hours. I knew this was my opportunity to wrap up the weather program to my advantage. I was positive I could pass the physical, written exam, and flight test to officially get my wings.

My hunch was correct. Captain Schoffield advised me to go for a commission but I politely refused because it would cost me money and add social responsibility that I thought was counterproductive. I suggested that I take the training while working the weather flight and earn my wings in grade with any instructor that would make the weather flights with me. My inward feelings were that Colonel John would view this approach with favor because of cost and expediency. Also, the weather flights would be uninterrupted. Captain Schoffield agreed and promised to present the plan to Colonel John and Captain Smith.

I was informed by Captain Smith that I should report to the flight surgeon on Thursday, 13 July 1933, at 1000 hours for the physical portion of my 64 exam — the Army Air Corps cadet exam for pilot trainees. If I passed, he would brief me on the written portion. There was no mention of a flight instructor.

The flight surgeon, Captain Jim Bowman, was no stranger to me because of the Fritzie encounter with the Boston Bull. On arrival at the base hospital, however, I had a cold and was tired from the morning flight. Doc took one look and challenged me: "You flew this morning with a cold?"

"Yes, Sir," I replied.

"How dumb can you get? I'll do the preliminaries and give you medication for it. You're grounded for three days. You should be okay after that. Check with me Saturday. We'll schedule your exam Tuesday, 18 July, at 1000 hours. I'll also check you for Monday's weather flight."

After a full day of rest on Wednesday and ten hours of sleep on Thursday I was in the pink. To pass the time I went over to the Signal Corps pigeon loft, the operation managed by Sergeant "Shorty" Weines, a close friend of Colonel John. Shorty had asked me on occasion to help in ground training, which involved taking a load of pigeons in the sidecar of the loft's motorcycle for measured distances from the loft, each trip a little farther away. It was fun to see the birds spot their direction for the home flight and then be back at the loft to see them return to their mate and their nest. Shorty would say, "This is real love."

Colonel John loved his pigeons and was a recognized pigeon racer. I did the right thing in helping Shorty. Colonel John frequently visited the loft, and he knew my interest in the training and would talk to me about its technique. I knew that I was making points with him, which wouldn't hurt in my desire to take over the weather recon program.

Saturday morning I reported to Captain Bowman at the hospital, for my re-check. He wasn't there, however his aide was on duty, and he checked me over and cleared me for the Monday weather flight.

That Tuesday, at 1000 hours, I reported to Doctor Bowman for my physical. It took the whole day. There was no food until supper, which I ate in the hospital. I was cleared to make the weather flight on Wednesday. Bowman ordered me to report to him Thursday at 1400 hours for the results of the physical. I knew that I would pass but was surprised that he ordered the weather airplanes to be equipped with oxygen as soon as possible. Doctor Bowman said, "You're showing some lung irritation and chronic nasal drip. You pass the test, however. If you do as well on your mental as you have on your physical, you'll be okay. I'll send my report to Captain Smith. Good luck!"

"Thank you, Sir." My response was very sincere.

Although I figured I would hear from Captain Smith by Monday, I was not sure how he would handle the written exam. The flight Friday was rough, with thunderstorms over the field and turbulence all the way up and down. Ben Dalley rode shotgun and wryly noted, "You'll take more crap than a Missouri mule to get a report and another notch on your gun. This is the last time for me in such conditions. I'm not going to bust my butt to feed your ego."

I was surprised at his remarks, but, in fairness, the flight conditions really had been bad. I told Ben, however, that if everything went as planned, I'd be on my own; he wouldn't need to fly with me again if all went well. Ben

The PT-3 Air Corps Primary Trainer used by the Army Air Corps in the late 1920s and early 1930s. The author was transitioned to the 0-19, flying this airplane.

ignored my comment and headed for the locker room.

On Saturday, I was in Operations checking on weather developments after the rough flight Friday. Captain Smith was signing a plan for a local flight. He saw me and asked if I wanted to take a ride. "Get your gear and come along." The airplane was a PT-3, and I realized he was going for some aerobatics — my cup of tea!

On the way to the airplane he said, "Take the rear seat and hook up the gosport" — the gosport was the tube connecting the pilot and copilot for communication. I suddenly knew what he had in mind; it was obvious that he wanted to check my ability as a pilot so that he would know at what level to start my training.

We took off and climbed to about 8,000 feet, and without a word to me, he stuck the nose down. At 120 mph he pulled up into a loop, nothing fancy. The next maneuver was a split S after pulling out of the loop dive. He climbed back to altitude and did a slow roll to the left, then slowed more and did a left snap roll. This was about the limit of the airplane's ability.

The Captain came over the gosport, "Take over and show me what you can do." I was determined to give him a good ride. I had my routine worked out from my practice flights with Ben Dalley. First, a three-turn spin to the right. Then, immediately after recovery, three turns to the left. After this, a dive and a tight loop followed with a half-loop with a half-roll to the right on top. I climbed back to 8,000 feet and did a slow roll to the right and one to the left. I could see Captain Smith's hand on each side of the front cockpit. This was a sign he was getting jostled enough to hang on. He came on the gosport, "That's enough. Take me home. If you can make a good three-point landing, I'll make believe you can fly."

"I'll take that as a compliment, Sir! Sorry I couldn't show you the rest of my routine." The landing was a good three-point at minimum speed, with just a little throttle before touchdown.

We walked into Operations. Little was said except Captain Smith remarked, "Your written will be after lunch Monday in the library. Sergeant Finley will be your monitor. This is not an open book exam." I thanked him and went into the weather office. Ben Dalley was waiting for the Captain to leave to tell me that he had been out in an 0-19 and saw most of the maneuvers. "Looked good."

I replied, "Ben, it was your stuff; but how did Smitty like it?"

"He didn't say," Ben remarked.

I was frustrated. "That figures! He hasn't been on his back for a long time."

Ben was pleased. I could tell by his grin. He asked, "What's with the examination?"

"Captain Smith has scheduled me for the cadet written on Monday," I replied.

"You'll have fun with that," Ben answered. "Good luck. See you later."

I was a little apprehensive about the written test, but Dutch and Ben had briefed me on what to expect so I had boned up on certain of my weak subjects. I went into the exam with a good attitude. Sergeant Finley had taken the exam for both aircraft and lighter-than-air and was a pilot in both. He had done just as I was trying to do.

Finley and I met Monday, the 24th of July, in the library at 1300 hours. To my surprise, we spent an hour going over the examination. The Sergeant asked if I had any questions, so I selected several items on the exam and he freely gave his comments. He obviously wanted me to pass. I could not believe a noncommissioned officer would be that lenient.

The examination took six hours. Afterwards, we went to the PX and had a hamburger and discussed my plans for the future. Finley was strictly career and tried to talk me into staying in. He believed we would be at war in less than ten years and that high-ranking enlisted personnel would be in an ideal spot over commissioned reserves. As it turned out, he was right. But as I look back, I wouldn't have changed my life one bit. Getting married later and having a family turned out to be the best thing that ever happened to me.

My exam grade came from Operations — an "A." Captain Schoffield and Colonel John both congratulated me. I was flattered, but immediately was wondering who would be my flight instructor.

Elton Rowley, meteorologist and pilot, 1933-1934.

Master Sergeant DePussen

It was two weeks before I was to report again to Captain Smith at the 15th Observation (OBS) headquarters. I was informed that an instructor pilot (IP) from Kelly Field would be reporting for detached service — not permanent — within ten days. He would give check rides to all second class pilots to upgrade to first class and give me my flight test and, hopefully, my wings.

The IP was an old timer, a Master Sergeant with over 20 years' service, who had started flying in World War I. His name was Pierre DePussen; he was a French Canadian who had been an instructor at Kelly Field for years and had earned the nickname "Poncho." He was a small wiry man; his weather-beaten face never had a smile. I had first met him in Operations after a weather flight. Sergeant Finley had introduced him, but I didn't like his looks or his general demeanor — he seemed overly pompous to me.

Finley was scheduled for a check-ride upgrade, so I figured I would ask later his opinion of our instructor.

Poncho knew that he was under scrutiny. I could sense this in Operations, overhearing his preflight briefings and his response to a simple "Good morning, Sergeant." He would never use your name; it was always your rank.

When Sergeant Finley finally took his ride with Poncho, I didn't see the beginning because it was in the early morning during a weather flight; but I was able to talk to Finley after my return. His comment was simple: "DePussen is tough but fair. He'll surprise you. Be ready for anything."

Poncho had a lifetime of experience in training students. I guess he had checked out every second-class pilot on base before he got to me. By that time his wife had arrived and had temporary base quarters, so as a courtesy I invited them to dinner at the NCO Club. Poncho refused, however, with the comment, “Thanks. After we fly, perhaps. By the way, your instruction starts at 0800 on Monday, 14 August. We’ll meet in Operations.”

I immediately replied to him, “Sergeant, I can’t accept that schedule because I have weather flights at that time, Monday through Friday.” Poncho swore under his breath and walked away.

I went directly to Captain Schoffield’s office to report my problem, but he assured me that there was some misunderstanding and he could clear the matter with Colonel John and Captain Smith. Schoffield added, “I have a feeling you don’t like DePussen.”

I replied, “I think it’s the other way, Sir. He has indicated displeasure in several ways about riding with me in the weather airplane. I feel he thinks he’s getting put on the spot as an IP.”

Weather was tough to fly — you had to do it right to get a good smooth trace; you had to be accurate with airspeed and climb rate. And though Poncho was an excellent pilot in general, he was inexperienced in weather flying.

Captain Schoffield replied, “Well, good for him. He might learn something. Don’t fret. I’ll straighten things out.” I left the Captain with the thought that Sergeant DePussen was going to get told what to do and when, and that I was going to get the flight check of my life. I was confident that I could handle anything he could dish out, either in an 0-19 or a PT-3, but I hoped that he would ride “shotgun” on a weather flight in really bad weather. It was my experience that this was a real test of a pilot’s proficiency — and I wanted to show him mine.

It was Friday the 18th before I heard from Captain Schoffield. He informed me that my first flight with Sergeant DePussen would be after 1300 hours on a date at the discretion of the Sergeant. I heard from Poncho that same day. I was to report to him in the pilots’ briefing room at the 15th OBS Squadron on Monday, 21 August, at 1300 hours. My moment of truth was at hand. I was finally going to get a chance at the brass ring I had dreamed about for so long.

Monday was a hot day. The weather flight was beautiful and, with oxygen, not tiring, so when I reported to Sergeant DePussen, I was relaxed. But not for long!

My first briefing from the Sergeant was a surprise and was designed to shake my confidence in my ability as a pilot. His remarks, as I remember, were, “You think you can fly. I want you to remember that I will make you forget everything you have learned and you will learn the military way.”

I thought to myself, “Sergeant DePussen doesn’t know my first flight instructor, Otto Enderton; he could cuss you out and make you like it!”

That first flight with Poncho was in a PT-3. He told me to take off and climb to 7,000 feet, do a two-turn spin to the right and the same to the left, and recover on the point, the blimp hangar on the field below. This was no challenge to me.

I already had been doing loops, spins, and rolls — unauthorized of course — with Dutch Kleinoder and Ben Dalley. Sergeant DePussen was asking me to do elementary maneuvers in which I was already proficient.

I executed Poncho's maneuvers from right to left by coming out of the right spin dive to an accelerated stall into the left spin. I undershot the point about 5 degrees. This got my first ass chewing! I knew then that nothing but perfection would satisfy Poncho. His next instruction was, "Climb back to 7,000 feet and do a left spin, three turns without the accelerated stall, and come out on the point! And I mean *on the point*! Is that clear?"

I replied, "Understand, Sergeant, but it would be better if you didn't shout in the gosport."

There was no reply. The spin was on the mark, but I must have unnerved him a bit by pulling into a tight loop on the dive recovery after the spin. I went ahead with snap rolls left and right and slow rolls left and right. I deliberately worked my way to the downwind side of the field to see if he would give me a dead-stick landing.

The Sergeant was too quiet, so I cut the engine mag switch from the rear seat and dove back toward the field with the prop windmilling. I turned the switch back on and the engine caught; the landing was normal. He came on over the gosport. "Take me back to the line." There was an ominous tone to his voice.

Sergeant DePussen jumped to the ground and stood by the rear cockpit looking up at me so damn mad he couldn't talk. I cavalierly grinned down at him. "Hell, Sergeant, can't you take a joke?" I patted the top of his head. I was frustrated and angry to be forced to do things I was already proficient at, but no doubt this would probably be the last straw for Poncho.

I began to feel bad because I was sure Poncho was going to bust me out of flying for good. He had walked away without saying a word — the old sweat-out routine. But it would be my word against his, and I was sure Captain Schoffield and Colonel John were on my side.

Captain Schoffield called me to his office on Thursday afternoon. I figured this was the end of my flying; but to my surprise, he was smiling when I walked in. "At ease. Have a seat. Well, how are you and Sergeant DePussen getting along?"

I replied, "Just fine. He wanted to retrain me the military way — by the book, but not taking into account my past training. I feel that's unnecessary because I was trained that way from the beginning. Frankly, we're having fun. We don't agree on everything, but if we did then one of us wouldn't be necessary! Sir, I would like to have the Sergeant fly with me on weather flights. After all, that's the reason for teaching me to fly the military way. So far, I'm at a loss to see the difference — it's stick and rudder as far as I'm concerned. Flying the weather takes a lot more skill as a pilot as well as knowing the proper flying technique to get a useable aerometeorograph trace."

Captain Schoffield thanked me for the briefing and commented, "Sergeant DePussen, I'm sure, wanted to check you out on standard flight procedures before he turns you loose on weather flights." There was no mention of DePussen's displeasure with my flying. The Captain dismissed me and I returned to the weather station to check on conditions for Friday's routine flight. Finley asked me to check in with Poncho at the 15th OBS, and I figured I was finally going to have it out with the Sergeant, that he would bust me, or else I was misreading his methods.

DePussen didn't receive me with open arms, but he was reasonable in his comments about my showing off doing maneuvers that were over my head, plus not waiting for orders from him. And from now on, he said, I was on notice to obey orders or suffer the consequences. He also told me to plan my dual cross-country flight of 400 miles and return. The course had to be triangular.

The cross-country was planned for the next day, Friday, 25 August, to Kansas City, Missouri; then to Wichita, Kansas, with a layover Saturday; and then back to Scott Field direct, with an alternate stop for fuel at Springfield, Missouri. I presented the plan to Poncho along with a weather program. He bought the idea to take one of the more powerful weather 0-19 biplanes and leave no later than 1300 hours Friday. I wanted to go to Wichita because of Lieutenant Kleinoder's and Shedd's plans to start an aircraft factory there.

Dutch had a design for a sport biplane and had found himself a backer — a financial "angel." He wanted me to be his liaison between the engineering of the aircraft and its manufacture. I told Kleinoder about my planned cross-country on Friday to Wichita and asked if we could meet there. He agreed, and I told Poncho. His reaction was as expected: "As long as you don't 'fly on' Kleinoder." Poncho wanted me to fly on my own. "Remember, this is a check on *your* navigation and general airmanship."

I talked with Kleinoder and he agreed that he and Shedd would leave Scott a few minutes after I did. They would land at Wichita's city airport on the old California section southeast of town.

The flight was great; the weather was good all the way. I met Kleinoder and Shedd, as planned, and went downtown by taxi to the Broadview Hotel with Paul Yankee, the "angel," from Wichita, a well-known financier and backer of aviation companies. We had dinner together and even included Poncho, who was flattered; I was happy to see him relax. I think the Sergeant was beginning to believe that I knew the score.

Our meeting to look at the proposed plant at the Yellow Cab hangar south of Beech Field was on Saturday. Mr. Yankee was very businesslike and critical of our plans to build a sport airplane in the middle of the Depression, and frankly he was not interested in investing money in anything until the economy was better. This was a blow to Kleinoder and Shedd, but we agreed to put the project on hold. I was disappointed, too; I liked Wichita and had hoped that someday I could live there and work at one of the aircraft plants.

Sunday morning we went to the airport, checked the weather, and prepared to leave. The weather was poor over the Ozarks — low ceilings and possible rain. Poncho agreed to try Chanute, Kansas, as an added alternate. I didn't agree with the decision, because the front was moving to the southeast; we could get on the back side of it if we would instead set our course for Jefferson City, Missouri, following Highway 54 from Fort Scott, Kansas.

This, after all, turned out to be the best, and Poncho allowed that I was correct when I stuck the nose into a Kansas rain, then went north and broke out into the clear. Southwest of Jefferson City, the wind shifted so that we didn't have to stop for fuel. From Jefferson City, we had a west-northwest wind all the way to Scott Field.

We landed at Scott, low on fuel, but safe; we actually had enough fuel to take us to Lambert Field in St. Louis, if necessary. The total flight time from Wichita was 3 hours 50 minutes. Poncho was forced to agree that we had made the right decision, and I think he realized that weather *experience* had a great deal to do with the success of the flight. However, he was not ready to turn me loose. There were many hours of hard flying ahead of me before I would get my wings.

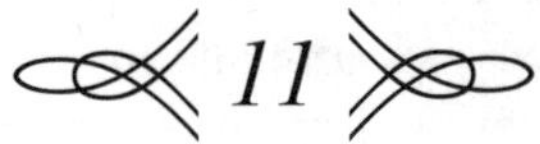

11

The Real Poncho

Sergeant DePussen had been flying with me for about 12 hours, and every flight was lint-picking from beginning to end. I had the feeling he was pressuring me for a reason, but was not sure why. I knew my flying was above average, but he would never give a compliment, only criticism. I was determined not to blow my cool. I was getting a lot of flying time with the weather run every morning and three or four hours of dual with him every week. This was more time than anyone else on base, so I was reluctant to ask for any change. The only problem was that my flight-training record would show more dual before solo than anyone in the history of the Signal Corps or Air Corps. I was taking a razzing for this from Ben Dalley and Dutch Kleinoder. My only response was, "Ponch and I are working up a lot of special training for you second class pilots just out of school." I didn't know how close this was to the truth; they would be going through similar "workouts" with Poncho in order to get their first-class rating.

However, the constant chewing from Poncho for no good reason was getting on my nerves, and the comments and razzing from Dutch and Ben about still being in training and not soloing didn't help either, so against my better judgment, I decided to approach Poncho man to man and, by force if necessary, settle the problems between us. The practice field near O'Fallon, just east of Scott Field, would be the perfect place because there was seldom anyone around.

It was a beautiful fall day. Poncho had scheduled a two-hour session in one of the PT-3s. This meant more practice landings, 180-degree overheads and dead-stick exercises, but I flew directly to the O'Fallon practice field.

Poncho was puffing over the gosport like a mad bull, something about that wasn't what he wanted. My response was swift: "Shut up you little bastard, this is going to be my day."

I landed and taxied to the downwind side of the field and cut the engine. He came out of the front cockpit, chute and all. I pulled him down from the wing, flat on his back. This brought fear in his eyes. I shouted at him, "Don't get up if you want to get away short of a meat wagon! Ponch, for once you are going to listen to what I have to say." I reviewed that I had flown dual for over 12 hours and hadn't soloed, had been called all the names in the book, and had never been given the courtesy of being told *how* to do things his way. (That *was* Poncho's way, however.)

"Now, Poncho, you're going to crank the engine and I'm going to fly this bird around the patch solo and make two landings. I'll then take you back to Scott Field and you and I will go to the Sergeant Major in Colonel Peglo's office and I'll turn myself in." Poncho cranked the engine and walked away and apparently watched. I taxied back after the landings and without a word he climbed aboard and we flew back to Scott.

We started walking toward Operations. To my surprise, the Sergeant stopped and extended his hand as a peaceful gesture. I refused, but did listen to his comment. "Congratulations. I wondered how long it would take to get a rise out of you. I don't want pantywaist students. Rowley, you've passed your test flight. All you lack is your solo cross-country." After the initial shock of his comment, I began to feel badly about my feelings toward him. Should I apologize, thank him, or just walk off and think the whole thing over lest I misread him again? There was one thing certain: My immaturity was showing. Poncho broke the silence by saying, "I'll be in touch tomorrow." I walked away without saying a word. I admit for once I was speechless!

My thoughts for the next few hours were mixed: what I had done wrong from the beginning and how Poncho had really gotten under my skin. One thing was clear. Who was I to try to out-think an old-time flight instructor who had, God knows, how many students with worse hangups than mine? My instincts, however, told me that there was more to come. Captain Schoffield called me to report to his office to review my flight-test record. I arrived with all the bad thoughts any young pilot could have. Captain Schoffield was in good humor. I was asked to be at ease. The Captain got up from his desk and shook my hand. The spell was broken when Poncho arrived. He was relaxed as hell with a half-grin on his face. He also shook hands with me. Captain Schoffield said, "Poncho, you've made another."

Poncho replied, "A good one, but it wasn't easy." It sounded like they were training horses.

I soon realized that I was getting a little kidding because it was obvious that Captain Schoffield and Poncho had discussed my training and experience in detail to the conclusion that Poncho wouldn't fly with me on weather flights because he had not had the experiences I had and thus he would be at a disadvantage as an instructor.

But the weather flying problem with Poncho was settled by the Captain informing me that this phase of my training was "special" and thus would be called "Special Advanced Procedural Flying," which Poncho, of course, was very equipped to handle. The training would start after my solo cross-country was completed. It was definite that I was going to get Poncho on a weather flight, even though we wouldn't actually call it that.

I went to work on the solo cross-country planning. The rules were strict, and the paperwork on the flight plans for each leg of the course was like a scorecard. The towns I picked were Terre Haute, Indiana; Chicago; and back to Scott Field. I was required to have at least two fuel and two weather alternates on each leg. At each main town I had to check in, refuel, and get the senior officer to verify my arrival time. The officers were National Guard types. I was watching the weather very carefully because I didn't want the embarrassment if I screwed up on my best subject; I was afraid I might never live it down because of the needling about some of the bad weather I had flown during our recon flights before I was authorized to fly them alone.

The great day finally came. The weather was perfect. I took the first leg to Terre Haute after the weather flight on Friday — same airplane, same clothes, short lunch; it was slow going because of a northeast wind. Two hours even, 160 miles, 30 mph head wind. I missed my Estimated Time of Arrival by five minutes. That night I stayed in the National Guard transient quarters. They were a nice bunch of guys; it was a lot of fun for them because they didn't get airplanes in very often.

I left Terre Haute at dawn. There was a light ground fog and very little wind. The flight to Chicago was 175 miles and beautiful. I realized as I approached the old Cicero Field, south of the city, that this was the first time I saw myself as a real future pilot. But one thing that reduced my anxiety was the Pratt & Whitney engine singing its beautiful song. Every pilot down deep has this thought: "Don't quit, baby."

I arrived at Chicago (Cicero) six minutes early, 1 hour 25 minutes from Terre Haute. It was 1000 hours — normally too late to be moving on — but I realized that with the 20 mph tail wind, I could make the 260 miles to Scott Field in less than two hours. I decided to have the engine checked and violate the overnight layover rule because if I waited to take off the next day, there was a good chance for a south wind of over 30 mph on my nose, which would put my fuel reserve in jeopardy.

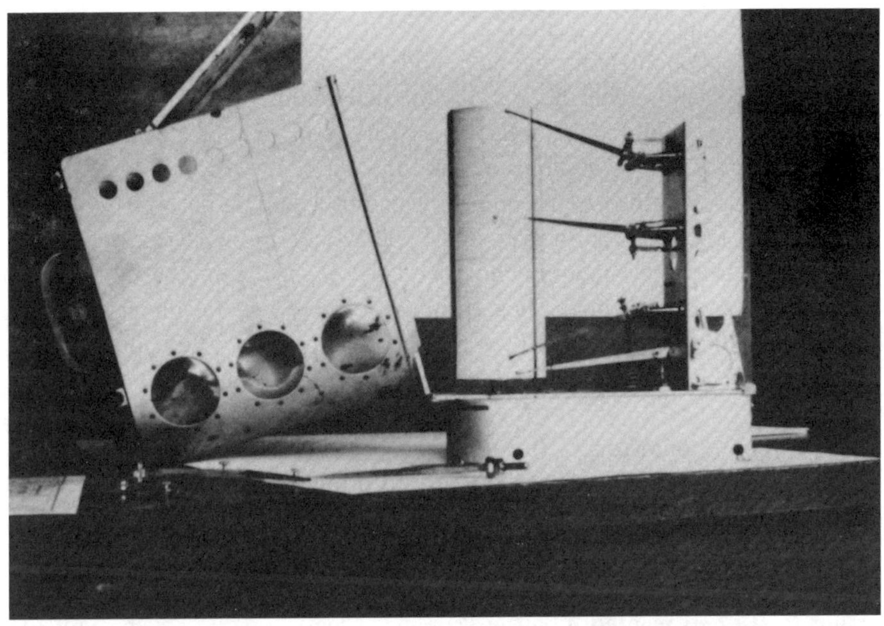

Above: An aerometeorograph, case open, showing the chart cylinder.

Below: The instrument case closed with mounting eyes showing on top. The lower mounting eyes are under the skirt of the instrument. The handle is removed for flight.

As it turned out, I was right. It was a beautiful flight. I arrived at Scott ten minutes early. Poncho confronted me with the overnight stay that I was supposed to have had in Chicago, and I explained that the weather was about to change and this would have made it necessary to layover past Monday; consequently, I would not be able to make the Monday weather flight out of Scott. Fortunately, this got me praise instead of a slap on the wrist.

My flight test was finally over, and I was wondering what was next. The weather flight came first. Poncho asked me to take him through all the preparation and procedures. It surprised me how well he understood the technical part of the program. I hadn't realized how much he had picked up on the side from Dutch and Ben. I asked him when he wanted to take the first ride.

"How about tomorrow?" Tuesday.

"Okay, Sarge. Dawn takeoff. Be sure to bring your face mask. You ride the front seat. You do the flying and I'll call the numbers." The last time I saw Poncho that day was with Jim Miner, our Crew Chief on the weather airplanes. Jim told me later that Poncho wanted to know which bird we were going to fly and wanted also the "squawk sheets" — the sheets with the pilots' mechanical complaints — on the bird for the last 30 days. This proved to me how thorough he was. My respect for Poncho was growing daily.

Tuesday was windy with a high cloud cover and a few breaks — a perfect day for Poncho's first weather flight: rough air, windy, a broken overcast at 10,000 feet, 7,000 under our maximum altitude of 17,000 feet. It was a little difficult to stay over the field because of the wind and the overcast, and the rough air made it difficult to make a smooth trace on the aerometeorograph. I commented to Poncho, "We should start our flight about three miles south of Scott to help on your drift problem. Also start your trace at best climb airspeed at each 1,000-foot level. Maintain your power off the power chart for each 1,000 feet of altitude." I was giving Poncho a little of the same crap he had been feeding me. Poncho was supposed to be the expert; I was the trainee.

We flew south of the field about three miles and started our climb. The first 10,000 feet was perfect, but when we hit the overcast, Poncho, to my surprise, said, "It's all yours from now on." I took over just as we entered the cloud layer. After another 3,000 feet, we broke out on top. I spotted the field through a break. We were a little too far north, so I corrected my wide spiral to get back on point within the 4,000 feet we had left in the flight. The end was over the field.

I touched the altitude pen using the control in the cockpit and flew straight and level five minutes to the south. Then with the carburetor heat on,

I put the airplane into a left spin to the top of the overcast below, about 9,500 feet, completed the spin recovery in the overcast, and broke out with my nose on the big hangar. With time and altitude to spare, I thought two or three downhill rolls would be in order. By this time Poncho's hands were hanging on the sides of the cockpit padding, which I decided was an indication to go home!

We took the weather instrument off the outer struts and went into the station; I knew Poncho was as anxious to see the instrument's trace as I was. The crowd gathered. Ben, Dutch, Sergeant Finley, and Captain Schoffield all were there. The trace was barely passable up to 10,000 — when Poncho was flying — and was real smooth after that — after I took over. The visitors razzed me. "You aren't so damned hot," they said. "Look what Sarge did on his first flight." Sarge started to explain.

I broke in and said to Ben and Dutch, "It can be done if you know how." Poncho smiled and let the whole thing drop; they didn't need to know that I did the last part of the flight.

I asked Poncho if he would like to see the read-out and the crowd began to leave, which really pleased him. He wanted to know why there was a difference in his trace up to 10,000 feet and mine later, in the overcast.

Always the instructor, Poncho finally concluded, "You must be very careful of your airspeed at the start of each climb level. The power setting to reach the climb rate is the trick." He further added, "This must be a real challenge when the weather is rough and you're flying blind." Poncho could clearly see what his problem had been.

I agreed with his evaluation and added, "It looks like we'll have one of those rough days tomorrow."

Poncho remarked, "Good, it's your day to fly."

The weather the next morning was far from perfect — ceiling of 1,500 feet, a 10 to 15 mph wind, and a cloud layer estimated to be about 3,000 feet thick. I had about three check points in the soup. The flight was real smooth.

Poncho asked, "Do you fly in stuff worse?"

"Oh, yes! How about being socked in all the way, up and down, and losing the field, hoping you can spot an opening and find your way home?"

Poncho said, "That's taking a big chance. What if you can't find a hole?"

"Well, Poncho, the last time that happened, I haven't found my way home yet."

He looked at me with a grin. "You're full of bull."

I smiled. "Maybe we'll have a chance to test our real skill just short of hitting the silk."

Poncho flew with me several additional times. He was very good and was proud of it. We had the usual bad weather, but he would take about as much of it as I would. Naturally, that cut down on the Go/No-Go hassle — the decision to fly or not.

One morning we returned from a rough flight. The trace was acceptable. Poncho looked at the read-out. He turned to me and said, "From now on you're on your own. You don't need me."

"Hell, Sarge, you can't do that. I'm planning on a 30-day leave and have told Colonel John that you would be the best man to take over while I'm gone." This really upset Poncho. He didn't want to fly weather. "What's more," I added, "I think you should transfer to Scott on regular flight duty. You and I could get checked out in the Ford Trimotor and the three-engine Fokker. We could get a lot of time flying cargo."

"I need flying time like I need the galloping dandruff," he answered.

"You say that," I replied, "but I'll bet your next month's pay that you have never been checked out in a Ford or a Fokker."

"You're right," he said, "I've been flying as a primary, basic, and advanced instructor."

"Sarge," I asked, "is that as far as you want to go? Of course, you're getting up in years. You have, however, weather recon experience. Why not get heavy birds under your belt before your 20-year retirement comes up? What's more, I'd like to ride copilot with you."

Poncho didn't laugh. I really believe he was starting to think a little about his future. One thing was quite clear. Poncho and I were working a lot closer and were trusting each other more as time went on.

Poncho had to leave; he had to go back to Kelly to settle his affairs with this new transfer to Scott. His final comment was very important to me: "You're checked out and you'll get your wings from Colonel Peglo very soon."

On Saturday, after Colonel John had called for a general inspection of the base, Captain Schoffield inspected the detached troops and as he passed my room he said, "Be in Colonel John's office at 1400 hours. Be prompt and shined and pressed. This is your day."

I was there five minutes early. The Sergeant Major asked me to be seated outside the conference room. The officers and Non Coms arrived promptly. Captain Schoffield, Poncho, Sergeant Bishop, Captain Smith, Major Koenig, First Lieutenant Holcomb, and Dutch Kleinoder came in and sat down beside me. Finally, the Colonel came out of his office and passed me without looking at me, his face very stern. I was beginning to think this was a court-martial. Dutch wouldn't say a word, but had a chicken grin on his face. We were finally called in. Dutch presented me like a new bride, "One pilot for your consideration, Sir."

Poncho handed me a glass of wine. Everybody stood up and Colonel John gave the toast, "May your troubles in flight be few and correctable before contacting the ground."

Colonel John presented me with my long-awaited wings. Second-class pilot, field grade. My speech was polite: "Thank you, gentlemen. I will do my best to serve both the Air Corps and Signal Corps as a soldier and a pilot. I would be remiss if I failed to mention Senior Master Sergeant DePussen in the expert way I was trained. Thank you, Sergeant, for your patience." I raised my glass. "Thank you all, gentlemen. Some day I hope I'll make you proud." This drew a loud "Hear! Hear!" Everyone shook my hand.

The Colonel closed the meeting. All my friends offered advice on continuing my career, but my mind was on that 30-day leave. I doubt if I heard half of what they said. I wanted to get home and make personal contact with my faithful girlfriend, Jean Austin. We had been corresponding since I had been in the service. And, of course, I wanted to see my mother and dad who had some doubts about my going into the service — particularly my flying.

Poncho agreed to stay at Scott Field during my leave, to do the weather flights. He also indicated that he might put in for transfer to the 15th. My furlough came through, to start 14 November through 18 December 1933.

Home on Leave

At about this time Dutch and Ben were due for their annual extended cross-country flights. I had been discussing the advantage of making them before December because of the unstable weather and suggested that they make their first leg from Scott Field to Cleveland, Ohio.

Ben said, "I'm for heading south."

Dutch's reaction was, "Hell, Cleveland is my old stomping ground. I'm for Cleveland."

Ben asked me, "When's your leave?"

I told him, "Next Tuesday, November 14th through December 18th." I added, "Poncho can fly the weather. I think I'll go along to Cleveland. I have a feeling Ben's bird is going to develop engine trouble." Ben's girlfriend lived in Columbus, and I was anticipating his hopes for a delay there.

Dutch said, "I'll call Pop Cleveland and we'll all have dinner at the Cleveland Hotel Wednesday night."

Pop was the manager for Cleveland Pneumatics, the company that built shock struts for aircraft manufacturers. Dutch had met Pop in Wichita when Dutch worked for Knoll Aircraft, before he had joined the service.

This was great for me because I had never met Pop, but had heard of and read several articles on the Cleveland air-hydraulic landing gear struts. I didn't realize how important Pop would be in my future flight-test career.

Ben decided to go along with the Cleveland scheme, but insisted that I had to do the lead flying.

"Hell, Ben, a blind jackass could find his way from Scott to Cleveland," I told him, "even by way of Columbus."

Ben insisted that I needed the practice. "If it's okay by Dutch, why don't you plan the whole cross-country, 800 miles plus. Get us out of the cold, like south of Cleveland. We want good weather and favorable winds all the way."

Dutch agreed, but added, "I'll bet you a sawbuck you can't work all my requirements into this flight plan."

"Fair enough," I replied. "I'll present the plan with dates for each leg. As to the wind, I'll check that off the weather station records. Favorable winds mean cross to directly on your tail. Let's shake on the bet before we leave."

Dutch remarked, "I have a feeling I shouldn't do this. I'll bet you've been working on the program for a week."

"Oh, you know I wouldn't do a thing like that," I said. "Particularly to one of my best friends."

"That's pure crap," he shot back. "You've beaten me for three times that much in poker. I wouldn't put it past Colonel John or Schoffield to be backing your roll."

I went to work on the flight plan for the extended cross-country after studying the weather and wind. There was a Pacific front approaching the West Coast, but after watching its movement, it appeared to be strong with a dry line. It wouldn't cause much temperature change or precipitation, and it probably would pass the Cleveland area during the night of the 15th and 16th, clearing out the morning of the 16th.

I had set the course from Scott Field to Columbus on 14 November: cross wind about 10 knots or less. The next leg would be Columbus to Cleveland to Raleigh, North Carolina: good wind, almost dead on the tail, north-northwest, 12 to 15 knots. Ben and Dutch didn't like this course because of the rough, hilly country and few fuel and emergency landing spots. This leg had to be flown on 16 November, as soon as the front had passed Raleigh. I suggested that they take off about noon to be sure the weather and wind were in their favor all the way. The longest leg was on 17 November, Raleigh to Memphis, Tennessee: cross wind at 4,000 to 5,000 feet; flight time 4½ hours, 675 miles, suggesting a fuel stop at Knoxville. The final leg was Memphis to Scott Field: 225 miles, 15 knots tail wind; a maximum 1½ hours chock to chock — start to stop — had to be flown on 18 November, midday.

I finished my plan and walked across the lawn to the BOQ Sunday, 12 November, at about 1000 hours. Ben and Dutch were both in the sack. I made some coffee and informed them that I was ready to go with the total flight plan. They both read and questioned every detail. I even suggested which airplanes to use — those with the low-time engines, the best they had.

Ben remarked, "Dutch, he's done all the work and don't forget he's going to fly my airplane to Cleveland. A lot is going to ride on his weather planning, say nothing of the wind, which is the key to the bet. Why not take the deal? You have a good chance to win. Particularly with the wind requirement. That's rough to predict."

We got off at 0800 Tuesday morning the 14th: light fog, cross wind southeast at 5 knots; beautiful and clear most of the flight. The time was 2 hours 45 minutes chock to chock. I arrived 5 minutes early; Dutch agreed that the first leg (Scott to Columbus) was a winner for me, though he called it "just dumb luck."

We all went to a beautiful Columbus residential hotel owned by the parents of Ben's girlfriend. We ate together in the main dining room, and Mr. Lawson announced that the dinner, breakfast, and quarters were on the house. I finished the meal, excused myself, and went to the airport to check the planes. I fueled and preflighted both birds, then went back to the hotel and hit the sack. I was sweating the front that was scheduled to pass during the night. Takeoff was scheduled for 1000. If the weather went as planned, we would have a northwest wind from Columbus to Cleveland — a cross wind.

My luck held; the wind was a dead cross. The second leg of our cross-country was completed, and we were in Cleveland on the 15th, where I was to leave the airplanes and continue on to Rochester by bus.

We had dinner with Pop Cleveland in the dining room of the old Cleveland Hotel, headquarters during the Cleveland National Air Races for racing pilots and "Quiet Birdmen" — the "Anciente and Secret Order" organization founded for aviators, in January 1921, by General Hap Arnold. It was a fine experience. Pop was a tall, handsome man, six-foot plus, with wavy gray hair and a large white mustache. He wore a tweed coat and whipcord (English worsted) trousers. He stood as straight as an arrow. He looked like a movie actor, but was easy to talk to and an all-around great guy. I had many fun-filled dinners with him over the coming years until his untimely death in Snoqualmie Pass, Washington, where Pop, his wife, and secretary all died in the crash of his single-engine Beech Bonanza.

On Friday, 17 November, I had breakfast with Ben and Dutch. We talked of our future plans in Wichita, then I boarded the bus for Rochester with my head filled with what was in store for me: marriage, a family, and flying. I was anxious to get home because I had been away for two years. I had kept in touch, but I missed the personal contact with my folks, and particularly, Jean. Our love affair, I was sure, was going to be permanent. Marriage was on my mind, but I couldn't get serious yet. I was due to finish my hitch on 11 July 1934 and Jean was planning to move to Milwaukee to work for Schusters, a large department store. My dreams of "tying the knot" were going to have to wait until I was situated in my flying career. The airlines looked good, as far as job openings, but seniority was a problem.

I arrived in Rochester at about 1900 hours and took the streetcar to my brother's home. I didn't phone; as a matter of fact, no one in the family knew I was coming. I went to the door in uniform, needing a shave, with a helmet and goggle tan line on my face, barracks bag in tow. Maurice, my brother, came to the door. He took one look, called my name, and hugged me like a long-lost friend. Father and Mother Swartout, Maurice's in-laws, welcomed me with open arms, as did Geraldine, his wife. Their seven-year-old son, Billy, was thrilled with the uniform; his three-year-old brother, Maurice, Jr., looked on wide-eyed.

Two years made a lot of difference in how people looked and felt toward each other. At first, I concluded that everyone was surprised at how I had matured; and the helmet and goggle tan line on my face made me look like an owl — maybe a wiser owl. I was 20 pounds heavier, no fat! Anyway, they seemed to be happy to see the young brother who wouldn't conform.

I called Jean and told her that I was in town and planned to stop on the way down to see the folks. My visit with Maurice was cut short, as I was going to take the bus and had to make its scheduled stop.

"Hell, Maurice," his father-in-law said, "let him have the pink Cadillac." It was a 1931, two years old. They always kept two. One old and one new. This was typical of Bill Swartout: a tough old man, but a heart of gold.

I left for the Austins, with thanks, in the pink Cadillac. But what was I going to say and how was I going to act on arrival? My fears, however, melted into pure pleasure. Jean's father met me at the door with a firm handshake and a slap on the back. Mother Austin (Louise) was next to greet me with a kiss. So far, so good. Where was Jean? Mother Austin broke the silence. "She'll be down in a minute." After small talk for about five minutes, Jean appeared. As they say down south, she sure was pretty! I casually walked across the room, picked her up, kissed her, and sat down with her on the sofa. I had my "weather eye" on Ralph and Louise. They laughed, and out came the chocolate cake and milk for me and coffee for everyone else. Jean wanted to know why my face looked so funny. "Lots of flying, sun, rain, cold, and sweat," was my answer. "Also a shave would help."

Everyone was interested in what I was doing. Was I flying every day? What kind of airplane? Could I keep warm at high altitude? How could I breathe? All the questions! I savored the opportunity to explain the dangers and the pleasure of my work. I proudly stated that flying and becoming a test pilot was my objective. Furthermore, I was going to be the best, because that was where the money was being made at this point in time. The reactions were mixed. I figured it was best to back off and change the subject, as I knew Jean would have something to say later.

Ralph joked about the pink Cadillac: "Be careful, don't stop, go directly home" — and so forth. We had fun. Then I headed out to see my folks.

I was greeted by Mother like she hadn't expected to ever see me again. Dad was at the general store playing cards, which was his habit. I talked Mother into going to the village to meet Dad and the other folks that were bound to be at the town's meeting spot. This was fun, and everyone had typical questions. Dr. McDowell, who had brought me into the world, looked at me and said, "I knew he was going to be different. Look what three years has done to that skinny kid that was jumping off the hill with his homemade flying machine."

The big question asked by most was, "Will you ever come back to Bristol Center to live?" My answer was a firm "No!"

Mother and Dad went on home and I followed soon after; I was ready to hit the sack. I thought about Ben and Dutch. They should be back at Scott the next day. I told Dad about the bet, and his remark was typical: "I think you thought it was a winner or you wouldn't have made the bet. I know you haven't got that kind of money to throw around."

The bed felt great, but at 4:00 a.m. I was wide awake and ready to fly. Wake-up time for the weather flight at Scott Field was 3:00. I soon dozed off, and Mother let me sleep until 8:00. I took a needed bath, shaved, and put on civilian clothes. Mother had bacon and eggs and all the trimmings. Dad had gone to the Sheriff's office; he was the Ontario County Deputy Sheriff. Mother and I spent the day talking and driving around the country. The weather was beautiful: cool, clear, mid-November. She had a million questions. Was I happy with my duty, did I have good quarters, and was I well fed? And what about Jean? Was I going to marry her? I answered, "I think I'm going to pop the question while I'm home. I plan to go see Uncle George, the jeweler, to get the ring soon." I assured her we couldn't get married until I was out of the service and until I had found meaningful work. This seemed to please her. She also said, "Jean is a nice girl and is from a very well-thought-of family. I hope you don't get mixed up with some unknown." I assured her this would not happen.

My leave was total pleasure: speeches at my prep school, chamber of commerce, and churches, and a meeting with my old flight instructor, Otto Enderton. We talked about the "fineness" of our trade — test pilots. He remarked, "You're on your way. Your service experience is great for selling your wares to the manufacturing industry." Otto was right.

We went flying together in a J-5 Waco, and I showed him what I had learned — the whole nine yards. He was pleased, and I felt that he was proud.

Of course, Frank Tones was on my frequent visitation list. And I worked on the pink Cadillac, tuning and washing down the engine, etc. I even found the airplane fuselage Clark and I had built years before with the engine and rudder and elevator installed and taxied it around on the fields nearby.

Thanksgiving was near. This meant that Maurice and Geraldine, the kids, and Mr. and Mrs. Swartout would all be at the farm. Christmas was always at Mr. and Mrs. Swartout's in Rochester. Jean wanted me to come to her home for dessert. I thought what a difference three years had made; I had suddenly become a hero in a small way among my family and their friends. My mother and I talked about this, and she remarked: "Don't let this attention go to your head. I have a feeling you have a long way to go." No words were more true, and I knew it.

Thanksgiving came and went with some sadness, as my time was slipping away fast. I had to get the engagement ring from Uncle George.

My uncle was a good judge of jewelry. He helped by selecting a beautiful yellow-gold ring with a pink Italian cameo. It was perfect for an engagement ring. On Friday, 24 November 1933, I spent the evening with Uncle George and Aunt Rosmond. They wanted to know my story about flying and Army life. George was my favorite uncle and my mother's youngest brother. He was always a good sounding board if there was any family trouble.

I had a date with Jean on Saturday evening for supper and intended to take a ride to McPherson Point on Conesus Lake. We had a pleasant evening dancing at the lake's pavilion and talking about our careers. On the way home, I asked her if she had thought about getting married: "I mean sometime in the future," I added.

"Yes," she answered, "but not until I have tried my retailing career."

We arrived at Jean's house quite late. I parked the pink Cadillac in the driveway and walked to the door with her. I took her hand and put the ring on her finger, kissed her, and said, "This is for real and forever." She didn't say anything. I walked to the car and left. I thought on the way home that I had heard of many ways to propose, but this had to be the dumbest.

On Sunday the folks took me to the old Methodist Church in Bristol Center. It had been built in 1850 and was small in present-day terms, but it was original and beautiful and had been my church since I was baptized. I'll admit, I was a little uneasy because of the uniform; the Methodists didn't believe in any kind of combat. The minister asked me if I was going to make the military my career. My answer: "I don't think so."

When we got home, the phone was ringing. It was Jean. She invited me for Sunday dinner plus a discussion about the unfinished business of last night. Mother said, "She has invited you for dinner again. It's about time she comes over here."

I said, "Mother, you set the time and I'll pass the word." Mother told me to invite Ralph, Louise, and Jean for 2:00 p.m. the following Sunday.

I kept my dinner date with Jean; I arrived at about 2:00. I felt I was going to have a real challenge, particularly with Ralph and Louise. However, to my surprise, they treated me as one of the family. Jean accepted my engagement ring, and we all talked about our future. I was relieved that the most important part of my trip home had been accomplished.

I was with Jean almost every evening for the remainder of my leave. Mother wouldn't let me sleep in, because I was keeping Jean up late. That was my mother!

Mother's dinner party was on Sunday, 3 December. The Austins arrived about 1:30. Mother was very quiet. Finally, Dad said, "Why don't we open up that vintage year Widmur Port and have a little toast to the kids?" I couldn't believe my ears. Mother was an absolute teetotaler. Without a word, she opened the bottle and poured about one ounce in each glass. I could see Dad and Ralph grin. Dad didn't comment to Mother about the amount. I think he figured to let well enough alone. Regardless of the volume of the drink, the feeling was there. Even though Mother didn't make a toast, her remark made the difference. "May God bless you both, and may your coming marriage be happy and fruitful."

I spent the remainder of my leave seeing old friends and spending as much time as possible with Jean. I visited Clark, who had a restaurant and motel, and I thought of how close we once were but now how different our lives had become.

Dad had made arrangements for me to ride back to Scott Field, St. Louis, with a New York state trooper who was traveling to his home near Bartlesville, Oklahoma. Trooper Mike Todd was an ex-rodeo and trick pistol shot — an interesting road companion. We left Rochester from my brother's, after I gave them all the news about Jean and me being engaged, which called for another round of talk and congratulations. I had tuned and washed the pink Cadillac for them before returning it. It was a beautiful car, but what a color!

Trooper Todd had picked me up at Maurice's at 10:00 a.m. on 16 December; we arrived at Scott Field late the afternoon of the 18th. Todd spent the night with Ben and Dutch, and I had a message to report to Captain Schoffield at 9:00, 19 December. All hell was about to break loose.

I had breakfast with Trooper Todd before he left for Oklahoma. He was a great guy. He resigned later and took over his family ranch and later struck oil all over the 2,000-acre spread.

The Air Mail Challenge

I reported to Captain Schoffield on 19 December, as ordered. He was informal, which meant that he had a problem. I asked how the weather flights were progressing, and he replied, "Fairly good. Poncho is getting better with experience. As a matter of fact, he will be here in a few minutes. You and Poncho are going to be involved in a very important program that is to start on 1 January."

Poncho arrived and the Captain started talking before Poncho sat down. Poncho and I exchanged handshakes; he was as wide-eyed as I was. Captain Schoffield proceeded to outline a program that had been assigned to the Air Corps. In our case, the 15th OBS Squadron was to fly the mail from Lambert Field, St. Louis, to Peoria, to Chicago, and return — the old Lindbergh run.

"I want you men to set up weather stations at each landing point, complete with teletype and maps. We're also trying to get homing devices and radio receivers installed in six airplanes."

This meant that our radio detachment would be involved, and we would need the Ford Trimotor or the Fokker to carry the men and equipment. We also would need the cooperation of the Bell telephone system to cut in the teletypes.

The Captain assigned the airplane and mechanics to Poncho and the weather stations and radio communications to me. We had about ten days to modify six 0-19 airplanes to fly one round-trip a day. We were to use National Guard facilities for hangar space, furniture, crew quarters, and food.

On 28 December 1933, Poncho and I had dropped off at each mail stop a mixed bag of weathermen, radio operators, and mechanics. On 31 December, we made our own check from Lambert Field to Chicago via Peoria. We had sent teletype messages to each station so that we could verify the weather there. The ceiling was about 1,000 feet with 3 miles visibility all the way. We were pleased with our check and stayed in Chicago to monitor the arrival of the first Air Corps mail flight.

By 1 January 1934 all weather stations were operating, and two airplanes were modified to carry mail sacks in the front cockpit. Radio use was by Code Word (CW) only. Short-range verbal receivers were being fabricated by our radio Signal Corps personnel.

Lieutenant Holcomb, the Vice Commander of the squadron, made the first run. He was a tired pilot, as the weather had deteriorated to almost impossible by the time he had arrived. He was on time, however, and mentioned that he had been warned properly of the weather conditions in Chicago by the weather station in Peoria.

Dutch Kleinoder was to take the flight back to St. Louis. Takeoff was scheduled for 1400 hours. The weather had improved very little. I told Dutch, "The 'iron compass' — the railroad — from Chicago to Peoria was good, but when you get down to the tops of the telegraph poles, land in the nearest frozen wheat field." He asked me if I would take the weather — would I fly in those conditions? I replied, "Sure. Those farmers en route have good-looking daughters!"

Dutch turned to Poncho and said, "How do you stand Rowley's crap?"

Poncho replied, "I've thought the same way. But I rode the weather with him a few times and I'll have to admit he can handle more than I'm comfortable with."

I replied, "Why not? Look, I've picked the best — Ben, Dutch, and Poncho. Dutch, if Holcomb will say the word, I'll take the flight." I admit that I put the pressure on Dutch. Lieutenant Holcomb overheard my comments and came down on me for this in his West Point officers' way.

I backed off. I asked the regular weather briefer to talk to Dutch and left the area. When I heard an engine start, I figured Dutch had decided to take the flight after all. But the weather was pure crap from Chicago to Peoria — the kind that produces dangerous rime ice. If it did start to freeze, Dutch would be in deep trouble; he had to stay just under the cloud cover — on the deck.

Poncho and I were going back to Scott Field as soon as the weather cleared. I had asked to have Peoria TWX — teletype — Chicago on Dutch's arrival, but the time had already passed for him to have landed there. I knew he'd had trouble, God knows where. In about five hours, we received the message that Dutch was down about 20 miles from Peoria. He had nosed over and needed a new prop. The mail had been taken to Washburn, Illinois. Dutch was okay; no injuries. We picked him up in Peoria the next day and flew him back to Scott. He was determined to make a complete flight from Chicago to St. Louis, so we flew him back to Peoria with a new prop. The mechanics were going to Washburn to do what was necessary to get the 0-19 back in the air.

While all this was transpiring, however, our weather recon flights had suffered to the point that Captain Schoffield asked me to resume flying the weather ship and keep an eye on the weather stations on weekends. Poncho and I would take the Ford Trimotor and make a round-trip on Saturday and Sunday. This was tough duty for me as it left me very little free time. Poncho was also busy rescuing downed airplanes.

As the 15th's air mail program moved on, up and down the St. Louis to Chicago circuit the flight record was very poor. The mail had previously been flown by the airline pilots, who were familiar with conditions and better equipped to handle them. The only good part was that no one from the 15th OBS had been killed.

Other flight sectors were very bad, as well, with several pilots lost. It seemed that from the East Coast westward to Chicago was the worst because of the lakes, though Chicago to San Francisco was almost as bad. On our "Lindbergh" run, we had flat country, and although the winter of 1934 was light on snow, it was very cold. This made pilot fatigue a real factor in the open-cockpit 0-19. The intense cold demanded better equipment, face masks, and better helmets and gloves.

I was involved with all of this because at the altitude required by the weather flight, conditions were below -10 degrees nine months of the year. Sergeant Bishop had been assigned to rework the existing clothing, install headphones in helmets that were insulated, and obtain thermal underwear and socks. Face masks were another problem because of condensation and subsequent freezing around the nose and lips. But I fixed this with a cotton undermask. The girls at the balloon (lighter-than-air) sewing room actually solved most of the problems; all they needed was a little direction, and they were off and running.

To further resolve the "air mail challenge," our radio shop had perfected for me a homing device — a loop antenna that would receive CW from the ground up to 20 miles. The loop was fixed so that the airplane had to be flown to the beam emanating from the ground station. The beam was north of the station and could be received by the airplane in good weather at a distance of about 25 miles. However, if there were any electrical storms in the area, static was a problem, and the range had to be cut to 10 miles or less. In the case of the weather airplane operation, I would try to keep the aircraft in a wide circle climb up to 17,000 feet. I could hear the beam "dit-da" (*A*) each time I passed north of the field.

To home in on the ground station beam required flying the aircraft north of the field until the beam was heard, then turning south 90 degrees. If the wind was favorable, it was easy to hold the beam. If the wind was from the right or left of the beam, it was necessary to zig-zag back and forth across the beam to judge the true heading required to return to base. The homing antenna could transmit and receive CW through the ceiling and bad visibility, allowing us to let down, to break out of the soup safely. This was a crude system, but it allowed me to make over 10 percent more flights in bad weather.

At this time, in addition, the 6th Signal Corps radio lab, in connection with the Signal Corps headquarters at Fort Monmouth, New Jersey, was working on a radial beam system that would transmit a different code signal on each side of the beam. Each station would also transmit a three-letter identification in Code Word.

This radio system was a great step forward and finally became the standard for weather flying for years. Not until World War II did we have ultra high frequency (UHF) and variable high frequency (VHF) radio systems that were static free, new approach systems, and automatic direction finders (ADF) that would operate on at least three frequency bands. The old beam system was used until 1950.

The beam navigation system was not perfected soon enough to help all of the pilots on the air mail caper, however. President Roosevelt gave the order to end the Air Corps assignment to carry the air mail in March 1934. We were to vacate all stations by 1 April. The result of the air mail experiment was a disaster in lost airplanes and pilots, but the overall effect was good. All services saw the inadequacy of our equipment and personnel training. General Benny Foulois, then Chief of the Air Corps, took immediate action by getting new airplanes on the drawing board and more money for equipment and instrumentation research. The net result, I think, was positive.

I continued to fly the weather until mid-May 1934, at which time I was assigned to the *Explorer I* balloon project as meteorologist and upper-air sounding expert.

I was back with Colonel John Peglo again, which pleased me. Colonel John called me to his office, where I met Lieutenants Stevenson and Anderson. I would enjoy this assignment as my last real experience in the service.

The *Explorer I* Project and Fox News Reel

The U.S. Army Air Corps had been working on the high-altitude research *Explorer I* balloon program for some time. Worldwide, the competition was intense for altitude research information. The Scott Field lighter-than-air operation was doing scale-model testing, preliminary to the actual *Explorer I* flight at Rapid City, South Dakota.

Colonel John called a briefing session on the *Explorer I* balloon stratosphere project on 14 May 1934 at 1300 hours. Those present were Major Kenney, Captain Schoffield, Sergeant Bishop, Sergeant Finley, Jim Scham (the civilian foreman of the balloon shop), and me (the project meteorologist). I made the weather flight that day on schedule and a balloon sounding after the flight. I knew that there would be questions about a "wind hole" in order to make a model trial flight, so I prepared myself the best that I could. (The wind hole was actually a spiral effect caused by the wind changing direction at different altitudes, creating a circular pattern. This "hole" allowed the balloon to go straight up and stay close over a point without drifting). The wind-hole questions would have to be answered before the scale-model balloon could be released. We had to be able to see the balloon for at least 35,000 feet to determine how well the bag design would take the gas expansion.

Colonel John laid down the ground rules. The scale model gondola was not to be pressurized, but would carry scale weight that included an aerometeorograph. Sergeant Bishop was made responsible for the inflation of the scale model balloon, and Jim Scham, along with Captain Schoffield, was to check instrument installation. I was to set the release time as soon as we had discovered an appropriate wind hole. Colonel John insisted that he was to see the wind-hole plots from the theodolite readings before the order to release.

Colonel John closed our first briefing by advising us that there would be Fox Movietone News personnel on base soon and we were not to discuss the project. Major Kenney was to be the "PR" officer and would be responsible for the overall coordination. The Fox people were in the area awaiting the arrival of General George Patton, on his cross-country trek to prove the mobility and value of mechanized warfare.

Captain Schoffield and I went back to his office. We went over the program in particular. He asked my opinion regarding the chances of finding the so-called wind hole, and my answer was, "Less than 50 percent at this time of the year. But there are two other options." That got his attention. I explained that I had been working on several ideas. The first plan was to locate an airplane at 20,000 feet with a long-range camera and an expert aerial photographer. The location would be pre-planned by weather-balloon soundings. As the scale-model *Explorer* balloon would pass the airplane, pictures could be taken at 5,000 feet below it to 15,000 feet above it. This plan would be flexible, because the airplane could maneuver to get better photos, to find out how the catenary ring at the top of the balloon responded to altitude. Twenty thousand feet was possible with one of our weather airplanes. I would fly the mission because I had had more experience at altitude than anyone on base. We also had the oxygen equipment and clothing to protect the crew at 20,000 feet, and I knew the 0-19 was capable because I had flown it to that altitude.

The Captain showed interest and wanted to know about the second plan — using the manned balloon. I explained that it was to be released at the same time as the scale model, and pictures could be taken from it at will. This plan would allow picture-altitude coordination. Both plans had advantages and disadvantages on balance. However, I felt that the first, using the airplane to take the pictures, was the best.

The Captain remarked, "I think your ideas have merit. Why don't you write them up in detail and I'll submit the information to Colonel John before the next meeting." I had hoped he would request this as I knew how Colonel John would react. My guess was that he would want both plans even though we did find a wind hole. Colonel John would want to put on a good show for Fox News.

Major Kenney called another meeting at 1300 on 17 May 1934. I was prepared and had turned over the alternate plans to Captain Schoffield the day before. I had also convinced Sergeant Farrell, our new weather station manager, to take balloon soundings around the clock to watch for a wind hole. Colonel John was already getting morning briefings on surrounding stations as well as on Scott Field. However, I wanted to distance myself from the wind-hole decision because I was convinced we would not find one under spring conditions. Colonel John knew when the ideal conditions would occur because he had made his biannual balloon flight during late July and August in a wind hole. He was proud of his ability to ascend in a free balloon for six hours and not get farther from home base than five miles.

The same personnel were present at our second meeting on the 17th. All elements were reported ready except for our release time. Captain Schoffield gave my alternate proposals and proudly said that they were planned and written by our upper-air specialist, "Rowley." Colonel John thanked the Captain and me for the good thinking. I knew that one of the meeting members was ticked off because I had had the idea and had turned the plan over to Captain Schoffield, and he knew that he wouldn't have a chance at criticizing the plan after the Colonel had put his stamp of approval on it.

I thought that it would be a good idea if Captain Schoffield proposed to Colonel John that Sergeant Bishop fly the tracker balloon, as Bishop was supposed to be the best lighter-than-air expert on base. This would give him a chance to really show his stuff. These thoughts were running through my mind as Colonel John informed us that Jim "Red" Lippish, the Fox reporter, was outside and would be introduced to the group by Major Kenney.

We were all presented to the very amiable and very sharp head of the Fox team. He planned to cover our test as well as the flight of *Explorer I* at Rapid City. Major Kenney reviewed our program and the duty assignment for each person and advised Lippish that he would be glad to answer any questions. I could see that this was not going to satisfy our reporter; he seemed to prefer more "hands on" from start to finish. And I guessed that he would want to participate in the actual operations and would have cameramen present at all times.

Major Kenney and Colonel John dismissed the whole team, but asked Lippish to remain. Captain Schoffield walked to the weather station with me and asked my opinion on how to handle the requests of the news cameramen. I replied, "Major Kenney should turn Lippish loose to watch and learn about the technical side, which could help him at Rapid City. Colonel John could edit the news release and censor the film. The people involved are smart enough to give out only the information on a 'need to know' basis. There should be no problem." I then went to Sergeant Farrell's office to check on the upper-air soundings. No wind hole. Upper wind at 30,000 feet was 40 knots from due west; southwest lower.

Colonel John came to the weather office to check on conditions. Farrell took the briefing. I made myself scarce by going up the ladder to the theodolite platform to watch Dick Crawford take another balloon sounding, but my stay on the platform was short. Sergeant Farrell called to advise me that Colonel John wanted me to report to his office on the double. To my surprise, no one was present except me. The Colonel wasted no time in asking if I would fly one of the weather airplanes to 20,000 feet with an aerial photographer — and how soon. My reply was, "Sir, as soon as Captain Schoffield can assign a photographer." I added, "We should fly the later model 0-19 with the high blower engine."

Colonel John remarked, "Tomorrow morning should be okay after 0900. Would you inform Captain Schoffield?"

"Yes, Sir," I said.

Captain Schoffield assigned Sergeant Pierce, the photo lab chief. His selection was excellent, as the Sergeant had high-altitude experience in aerial mapping, though he had never used oxygen. That problem could be solved, however, with an evening practice session. I contacted Poncho late that night about the program; the Colonel wanted to start the next day. I proposed that I take the weather flight to check out conditions. Poncho agreed and mumbled something about telling Captain Schoffield.

The morning of 18 May was beautiful — not a cloud, and a sunrise as red as fire. The flight would be routine except that the visibility at altitude showed signs of dust. This worried me because good pictures of the model balloon might be difficult, to say nothing of the *Explorer I* flight from Rapid City. I also noted that our upper-air wind was strong from the southwest, which explained the upper-air dust. It was obvious that we were getting our problem directly from the dust bowl areas in Kansas and Oklahoma. The aircraft was fine. No squawks, routine service, and she was ready for the 0900 flight.

Pierce arrived at Operations about 0830. He brought a long-lens, shielded Leica. It was a little large to handle in the rear cockpit, but was workable with a safety strap. Colonel John arrived while we were suiting up; Lippish and Major Kenney arrived a few minutes later. By the time we were ready to take off, there was a collection of people, including Ben Dalley, Dutch Kleinoder, and Poncho. I could see in the faces of Ben and Dutch a combination of pride and envy. They walked out to the airplane as the Sergeant and I were about to "mount up," shook my hand, and said, "Don't screw up."

I replied, "Eat your heart out." They both grinned and gave me the finger.

The Sergeant asked me as we were climbing out, "Were those guys your friends?"

"You bet. They're my fathers. You're in my very young hands," I answered.

Pierce said, "How did I get this assignment?"

The climb was routine. The Sergeant remarked about the dust haze at

15,000 feet. It was steadily getting worse as we continued our climb to 20,000. You could barely see the ground. We still had about 200 feet per minute climb when the Sarge wanted down. We were just over 20,000 feet. We could have gone another 1,000, but I think Sergeant Pierce's oxygen was bothering him. The flight was routine except for the dust. I believed that we had obtained some good pictures with a lens filter; we would soon find out.

When we walked into Operations, Colonel John, Lippish, and Major Kenney were in the weather office, apparently comparing wind runs of the last few days. The Colonel was explaining to Lippish about the wind hole that was necessary before he could launch the model balloon. Major Kenney asked how the flight went.

"Routine, Sir. Dust was heavy at altitude."

Colonel John came out and commented, "No wind hole. Did you get pictures?"

"Yes, Sir," I said. "We will know by 1200 hours. As to the wind, strong southwest to west all the way. I have the aerometeorograph in the instrument room. Specialist Crawford is taking off the data."

The Colonel replied, "That was good thinking."

"Thank you, Sir," I answered.

The pictures turned out fair. The silver balloon would show up well against the ground and would be better against the sky as long as we could keep the subject down sun. This could be accomplished with the airplane.

Our deadline for the scale-model test was 5 June 1934. We continued to search for the wind hole. It became obvious we were not going to find the proper conditions to release the balloon by itself. It appeared possible, however, to find lower winds aloft about one day after each cold front passage. These conditions would be acceptable for our alternate plans, for either the tracker balloon or the airplane. Colonel John had to make a decision soon. As expected, he called the big meeting on Friday, 1 June 1934. All personnel were present. The Colonel had made the decision to use the tracker balloon *and* the airplane. The crew of the balloon was Colonel John, Sergeant Bishop, and Lippish as cameraman. Pierce and I would crew the airplane. I was assigned to follow the upper-air soundings to find the ideal wind conditions. Captain Schoffield was pleased that the Signal Corps plans were to be used.

A well-defined Pacific cold front had come ashore in the Washington and Oregon area, and it appeared to be the one we were waiting for. Sergeant Farrell estimated that it would pass Scott Field the night of 3 June. As the front passed, we had high surface winds, rain, and hail. The chances were bleak at about 0400 hours, 4 June. However, by 1400 the wind was dropping and the sun was breaking through. Farrell was ready to call Major Kenney.

I said, "Hell, no! I'll call Colonel John when we're sure." Farrell agreed that this was proper.

By 1800 we took a wind sounding in the blue. We were blessed with north and northwest winds below 20 knots at 25,000 feet; and at 35,000, it was 10 knots lower. I called the Colonel, and his only remark was, "Wunderbar," and he hung up. In seconds, I looked out the window and saw him coming across the street from headquarters. He entered the weather station in high spirits; he was really pleased and relieved. His remark to me was, "Let's fly at 0800 tomorrow."

"Yes, Sir," I said. "You can inflate outside early tomorrow morning."

The Colonel looked at me with a grin and said, "I see you have become a balloon expert already."

"No, Sir," I said, "just making a comment."

Pierce and I planned to be at 20,000 feet over the field at 0745. Our last balloon sounding before the flights had been at 0400. The results were good; the wind was steady from the northwest at 8 knots at 30,000 to 35,000 feet. The Colonel gave the word to go!

Sergeant Pierce and I took off at 0630. We had an aerometeorograph aboard and planned to touch the recorder pen each time we took a picture. This would allow a remote connection with picture-altitude. We were in position at 0745. The sky was clear and because of the upper-air wind, the dust problem was minimal.

Our radio operator gave me the countdown on the balloon release. We got the word at exactly 0800. I touched the aerometeorograph recorder pen. We were in an excellent position — down sun and about three miles southeast of the release point. Pierce started to take pictures every three minutes with an accompanying recorder pen touch. The balloons were flying well. The tracker was above and just northwest of the model, thanks to the great airmanship of Colonel John. I maneuvered the airplane so that the model balloon passed the airplane about 200 yards to the northwest. The picture was beautiful! We could get both in one frame. The model was taking shape very well. The inflation catenary was filling out as planned. The load catenary was hanging loose, but was not fouled.

We continued to take pictures as long as we could see the shape. Colonel John had started his descent, and I followed until we could estimate his landing point, which was near Addieville, Illinois, about 27 miles from takeoff. I radioed this information to Operations and was informed that an airplane had been dispatched to the area.

We had to land as we were getting low on fuel; our flight time total was about three hours. I went to the radio room immediately to find out about Colonel John. He had landed in an open field without any apparent damage.

Pierce and I went to the mess hall for an early lunch, then waited for the Colonel. Major Kenney, I knew, would not want to discuss the program until Colonel John and Jim Lippish returned.

Sergeant Pierce took the camera to the lab; he would have the pictures in about three hours. I had hoped that we could have them developed by the time Colonel John arrived back at Scott Field. I checked the radio station for any news; the operator said that the recovery truck had headed southeast on Route 177. This was perfect because Addieville was on Illinois 177. No other information was available.

The Colonel and Lippish returned to base aboard the Colonel's staff car at about 1800 hours. I was sitting outside his office with Major Kenney when he arrived. I had 24 beautiful shots, 8½ x 11 to present to him; they ranged from three minutes after release to approximately 30,000 feet altitude. The movies would be available later in the evening. I left the pictures with Colonel John, excused myself, and went back to the weather station to look at the recorder trace of the flight. I was sure that the Colonel would ask for a written report, so I informed Sergeant Farrell and he agreed to write up the information from the theodolite and the aerometeorograph. I agreed to summarize the information to present to the Colonel.

The expected call from the Colonel came. "Can you have your report ready by 0900?" he asked. "Captain Stevenson will fly in at about 1000 tomorrow. I'd like to go over all reports before his arrival."

"Yes, Sir," I responded.

It was near midnight when I completed my write-up and the Sergeant had it typed. I was surprised to find Colonel John still in his office. He asked me to stay and go over my report with him. Our flight pictures and the summary were complete; the Colonel liked what he saw.

I suggested to the Colonel that it had been a long day and that he should get some rest.

He agreed and asked, "Why don't you walk with me to my quarters?" He had a reason. I soon learned that he wanted me to stay in the service and continue doing the weather flights. He also as much as promised me a commission within three months. He continued to compliment me on the way I had performed my duty while assigned to Scott Field.

I replied, "Sir, I'm flattered and pleased that you approve of my work, and I've weighed my decision carefully. If you, Sir, were not retiring soon, I would stay. You've been very patient with me and have helped me through my training period. However, I must continue on with my objective to be a finished test pilot in civilian life. I feel I have the technical basis for that kind of work."

The Colonel listened and said he understood.

On the way back to my barracks, my thoughts drifted to Milwaukee and my Jean. Only 35 days until discharge. I hit the bunk, and the next thing I remembered was the reveille bugle. I made it late to breakfast, but the mess Sergeant was a good friend. We had eaten together many times at 0300, and he had the cook whip up a dandy for me this time. We went into his private area and talked about my future. Sergeant Price had seen many young men like me in his over 20 years in the service.

I was beginning to realize that the pressure was off. The flying weather was beautiful, so I wandered over to the pigeon loft where Sarge was training a new covey off the flight deck. The old birds were showing the young how it was done. I could understand why he liked his work. From there, I went to Operations. Ben Dalley was suiting up for a flight in a PT-3. This meant aerobatics. He asked me to come along, but I told him, "Sorry, the Colonel's having a meeting. Maybe tomorrow." I had a feeling something was "pulling my string." I would miss all this.

All were present at our final meeting on the *Explorer I* project. Colonel John reviewed his findings with Captain Stevenson. He used the airplane pictures to make his point regarding the gas-expansion effect on the inflation catenary: that the catenary band at the top of the bag was not allowing for expansion of the gas with altitude. The band was too small, causing a crease around the top of the bag. The Colonel flatly told the Captain that this was, in his opinion, a source of possible failure at extreme altitude. The Captain agreed, but thought that this could be a model effect and that the full-scale balloon might not react the same way. The Captain went on to state that this finding would call for an over-inflation test on the ground before the actual flight. I noticed that Mr. Lippish was taking notes on what was being said.

Captain Schoffield and I left the meeting after a quick dismissal of all involved. We walked toward the weather station as a matter of habit. It seemed that the Captain had something on his mind; he wanted to talk about my future in the service. His lecture was about the same as Colonel John's, and I responded in the same way. I promised to give it some thought and would report to him after my short leave to Milwaukee. I frankly told him that I wanted to discuss my future plans with my fiancée, who had a good job there. The Captain added that there were openings in the U.S. Weather Bureau if I wanted to leave the service: "I think you should give the matter some very serious thought." I thanked him.

Sergeant Farrell told me after the Captain left that he was going to Fort Sheridan on detached service with the 61st Coast Artillery. He asked me to go with him on my way to Milwaukee, to show him around, because I had been assigned the same duty two years before. I agreed to ride with him and stop off there to introduce Chief Master Sergeant Barrett, whom I knew.

Then I left Fort Sheridan and went to Milwaukee on the commuter train. I called my fiancée and we met that evening for dinner.

Jean and I thoroughly discussed our mutual plans, then agreed to get married as soon as I was established. We spent the weekend together. The decision to leave the service was firm. I returned to Scott Field by bus.

16

The Great George Patton

My last 15 days in the service appeared, at first, to be uneventful. My weather flying was over. The *Explorer I* project was successfully completed, and I would have no other assignments. This was the end of the military pressure, but I began to wonder what the challenge in civilian life was going to be like. By habit, I passed the days going to the weather station to take the morning weather briefing. I guess because of the bantering with Colonel John and watching Poncho work the weather airplane, I hassled Poncho about bad traces. His response was, "Why don't you stick around and do the damned flying yourself?" Frankly, Poncho's flying was barely acceptable, and he knew it. But I promised myself I wouldn't needle him further.

One morning about 0900, I was with Shorty watching the pigeons when Jim Lippish drove up in a news reel camera truck. I thought he had already left for Chicago, but he wanted to talk with me about driving the camera truck during General George Patton's 1st Battalion Mechanized Cavalry "foray" across country. Patton and his troops were to bivouac for two days at Scott Field and go from there to Fort Sill, Oklahoma, nonstop. My answer was, "Yes, if Captain Schoffield will approve." We went to the Captain's office and received immediate agreement with the added request that he ride along to watch the fun.

General Patton arrived at 1400 hours, 3 July 1934. He was riding in his specially built staff car, standing in a circular bar arrangement waist high, with steel helmet, goggles, pegged pants, boots, shirt, and tie. He directed his driver to stop in front of our headquarters. I was surprised that Colonel John didn't meet him at the door, but in less than five minutes, the Colonel's staff car pulled in front of General Patton's vehicle. The General and

Colonel John headed to the north side of the field in the Colonel's staff car. Jim Lippish was taking pictures from the weather tower. I was watching through the theodolite. You could feel the tension in the air. All airplanes were lined up; fuel trucks and any civilian vehicles were parked out of sight. The information and instructions about General Patton's visit had been passed on two days before, so we all looked sharp.

Colonel John's staff car was on its way back; I could see it through the theodolite. Jim and I both went to Operations to see the General and to hear the orders; Sergeants Bishop and Finley were behind the work counter. When the Colonel and General Patton arrived, we all came to attention, and I mean a brace. What a sight! I'll never forget. Colonel John was relaxed, but respectful. The Colonel's first words were, "At ease!"

Patton said loud and clear, "Thank you, gentlemen." The General was a picture of military elegance, both in stature and demeanor.

The orders given to Bishop and Finley by Colonel John were, "Advise all personnel involved that a special briefing will be scheduled at 1400 hours, in the theater, tomorrow, July 4th." Jim Lippish approached the General and stated his name and assignment. The General said, "Oh, yes. You will work with my information officer, Major Kelley." Lippish introduced me as his camera-truck driver. General Patton remarked, "I will assign a man to ride with you."

The 1st Battalion Mechanized Cavalry was standing at the northwest side of the field. The latrine and showers were the first to be assembled. There were about 100 vehicles and 250 men. The encampment was completed in less than one hour. Chow was served, line style. Non Coms and enlisted men ate in their vehicles, the officers in open-sided tents. Patton's headquarters tent included dining and sleeping with shower and latrine. The men and officers chowed down by 1830 hours. Patton was with Colonel John in his quarters.

At the end of my boot training in 1931, we had had one 30-mile march with all gear. I could, therefore, admire the precision of General Patton's 1st Battalion. It worked like a clock. Every element of the group knew its place and acted accordingly. Every man was razor sharp.

The next day, all Scott Field personnel were anxious to hear General Patton's briefing. We were all standing by even though it was "firecracker" day. The briefing turned out to be for the Fox Movietone News team and for the assignment of various key personnel to make the picture-taking well coordinated so that the column could maintain its nonstop plan and speed average of 45 mph. The final comment from the General was, "Our schedule will be lights out at 0900 tonight and reveille at 0400 tomorrow. Chow and break camp and move out 0600. You will follow our plan that you have in writing. Major Kelley will direct the camera teams, Fox and our own. Further, we will not compromise our mission for a picture record." Sergeant Baxter was assigned to my vehicle, and Major Kelley with Jim Lippish.

The Fox crew was in position at the encampment at 0500, 5 July. We were ready as soon as we had enough light, which was 0530. We were briefed by Major Kelley in detail on roads and routing through towns. Major Kelley's suggestions were helpful as to camera location in the center of the road and overpasses. We also had good maps to further help as we proceeded along a very difficult course. At about the same time that we started taking pictures, Colonel John, Major Kenney, Captain Schoffield, and Captain Smith paid their respects to General Patton. Captain Schoffield advised Lippish and me that he had decided not to go with us. I was relieved.

We covered the final breakdown of the camp, even to the filling of the latrine, then moved to the south gate to plan the movement from the base. The General was in his special vehicle, hard helmet and goggles, standing looking ahead, his right arm extended, waving forward. The column immediately picked up speed with about 50 feet between vehicles. The General and his column were a magnificent sight! Little did we know how important General Patton would be in World War II — his brilliant victories over Rommel, the Nazi's best tank commander and tactician, and his brilliant thinking over Britain's best, General Montgomery. That day, 5 July 1934, we witnessed the beginning of what has become a standard in all nations for a mechanized fighting force. (Patton is credited with developing the science of mechanized warfare.)

The weather was excellent — clear and warm. The camera vehicle was a modified Packard touring car — big and heavy: no power steering or automatic transmission; a large steering wheel; a three-speed stick shift; and no syncro gears. Pure truck, double-clutching! I was glad I had driven a heavy truck for my brother. You have to develop a technique to shift fast, and fast was the order of that day, from 75 mph to stop, all day long. The Packard was a powerhouse. The big in-line eight-cylinder engine was as smooth as glass. We could go to 60 mph in second gear and to 100 mph in third. Sergeant Baxter was a real pro at driving military equipment; he was a great help. He could anticipate the General's moves and joked about the old man's temperament and showmanship; but he was not disrespectful. It was obvious that he loved the General's ways. He had been with him about 15 years. Baxter had been part of the mechanization effort, both in design and testing. He had 20 years in the service and was one of the General's most trusted enlisted aides.

We scanned the General and the first half of the column and raced past to park at the juncture of the base road and the highway to Belleville, Illinois.

The Army photo team led the way. Jim Lippish didn't like this; he knew they would park in the choice spot. He remarked to me, "Next time this happens, don't let them pass."

Sergeant Baxter heard the remark and firmly replied, "I wouldn't do that. If the General sees such actions, you'll be on your way back to Chicago. Furthermore, you know it was agreed to coordinate all movements with Major Kelley." Lippish muttered under his breath.

Our course was Scott Field to Belleville, Illinois; St. Louis to Jeff City, Springfield, and Neosho, Missouri; and to Ponca City, Ardmore, Lawton, and Fort Sill, Oklahoma: a total of approximately 700 miles. Time: 16 hours. We arrived at our bivouac area at 2200, 5 July. General Patton was pleased with the performance of his group. They had made their speed average, as planned.

There was only one vehicle failure, and Fox lost one low-profile movie camera that had been placed in the road. After the camp was in order and our film had been canned, the General invited Major Kelley, Baxter, Lippish, and me for a drink and to sample some of the new field rations. The drink was Wild Turkey, straight and warm. It burned like fire, but put life into everyone. Lippish was not pleased because the footage had to be censored the next day, but the General made it clear as we were sitting around drinking and eating that he would personally review the movies the first thing in the morning, 6 July. Furthermore, there would be no unnecessary delay. General Patton also informed Major Kelley that we were to be housed in the Fort Sill VIP quarters and that we were to eat in the officers' open mess. As we were leaving, the General said to me, "I like your jump suit. I'm curious. Is it issue?"

I replied, "Yes, Sir. Air Corps new summer flying suit."

He asked, "What is your rank?"

"I have none, Sir," I said. "Specialist pilot, high-altitude recon for five more days."

"You did a fine job driving."

"Thank you, Sir, and to Sergeant Baxter." I was glad the General dismissed me, as I was out of uniform. I mentioned the weather: "Sir, by 1300 tomorrow, I predict thunderstorms and possible tornadoes. This is 'Tornado Alley,' Sir."

The General looked at Major Kelley and said, "Check further on this, Major. This could be real trouble and a hell of a way to end up a fine run."

As we drove to the VIP barracks, Major Kelley asked me how sure I was of my prediction. I gave him my best guess of 90 percent. Major Kelley headed for the radio room at base headquarters to check. I saw him the next morning in the projection and planning room with the General. After the film review, General Patton and Major Kelley asked to see me in one of the adjoining conference rooms. There was a large U.S. planning chart on the wall. The General pinned me down as to the storm track and time. I explained the frontal movement and the approximate position and origin of the low-pressure trough and its movement that had been predicted 4 July. If it continued on course, it would arrive in the Fort Sill area on 6 July, at approximately 1300 hours. Major Kelley explained that this was very close to the Washington Weather Center's forecast. The General gave the order to take standard severe wind and rain camp-breakdown precautions.

My prediction was very close — winds 45 mph gusting to 60 mph with heavy rain. A tornado touched down on the south side of Lawton, Oklahoma, five miles south of Fort Sill. I was proud to have been helpful to the General. And on his future travels, he would have a weatherman, even during World War II, which would prove to be one of the reasons for his great success.

Jim Lippish finished his film work and advised me of his plans to ship the cans from Oklahoma City's express office. Then we would head for Scott Field. My time was getting very short. Lippish tried to talk me into working for him as a pilot and driver for his team that headquartered in Chicago. Our first assignment would be the *Explorer I* balloon flight from Rapid City. This certainly sounded good, but it had no future, so I turned it down. Lippish stayed overnight at Scott Field on 8 July. I was discharged the 11th and planned to leave for Milwaukee on the 13th. This would be the beginning of a very different life, but was another notch toward my training as a test pilot.

Civilian Life in Milwaukee

As I prepared to leave Scott Field on 13 July 1934, I paid my respects to the officers who had helped me along the way: Captain Schoffield, Major Kenney, and Lieutenants Dalley, Kleinoder, and Holcomb. It was particularly hard to say goodbye to Colonel John. He shook my hand and put his other hand on my shoulder. His comments were brief: "Good luck. I know you have what it takes to be a good test pilot."

I sensed a little tremor in his voice. I had a few tears in my own eyes. About all I could say was, "Thanks for your help, Sir. I shall never forget. God bless you and Mrs. Peglo in your retirement."

Fritzie was in the office. I picked her up and told her to be good to the Colonel. She responded with a big wet kiss on my cheek. The Colonel smiled.

All the other officers and men were complimentary, including Poncho. "I'll be following you soon. I plan to retire in Texas. Let's stay in touch."

The weather station personnel, Sergeant Farrell, Crawford — my replacement, Dixon, Jones, and Sergeants Bishop and Finley were in the barracks when I returned after taking my physical. They had planned a small party in the detachment ready room — beer and some "rot-gut." The plan was for a little poker party. Everyone knew I had my mustering-out pay. I had a polite drink but declined the poker game. Bishop made a nasty remark about "no guts" and I knocked him on his butt. I picked him up, then reminded him that I had no rank. I was a civilian. This put a sour note on our get-together, but it was offset by Ben and Dutch inviting me to the Officers Club for dinner.

Ben and Dutch were running out of time, too. Their active duty was terminating in September. They wanted to know what I was going to do. I said, "I'm going to Milwaukee, get married, and rent the old Curtiss-Wright airport." I had wanted to try a fixed-base operation with a weather contract, good maintenance, storage facilities, and flight instructions. And I had a good opportunity to get all the necessary Bureau of Air Commerce licenses from the area supervisor, Ora Young, who was a friend of my original flight instructor, Otto Enderton.

Money would be no problem as my old New York Life Insurance policy that my dad had taken out for me when I was five would be good for a nice bundle. I offered Ben and Dutch space to build the prototype of their Wichita bird. All this got their attention. It was midnight before we broke up, and we agreed to keep in touch.

Ben was going to stay with his folks in St. Louis for the weekend, so he gave me a lift the next day to the downtown St. Louis bus station. His last words to me were, "Pard, be careful. Those civilian crates are not as good as the military. Knock 'em dead and good luck." He waved as he drove off. The sad part is that I never saw Ben again. He was killed somewhere in the China Sea in a MATS transport in early 1944.

My ride to Milwaukee gave me time to think. I had no job, $450 separation pay, and a girlfriend who was the greatest. And my airport dream was just that — a dream. The weather contract was a possibility, but I had to get a suitable airplane. The airport was occupied, so I would have to buy the lease. These thoughts were racing through my head. The solutions would be tough, but not unobtainable.

I arrived in Milwaukee at about 10:00 p.m. on 13 July and immediately called Jean. I had planned to stay at the downtown YMCA. It was a strange feeling to be just another civilian. In civilian clothes, I didn't feel properly dressed; but the clothes, I soon found, would be the least of my problems.

The YMCA director gave me a lot of help, along with some good advice, particularly on the availability of jobs that would fit my experience. After a few weeks, the best I could find was selling Eureka vacuum cleaners. This was, in many ways, good experience. The city was made up primarily of Germans on the north side and Poles on the south. My job was to be the demonstrator. It seemed that Saturday and Sunday were always booked solid from early evening to late. It was fun, but I soon realized that the people were using me for their entertainment. The neighbors all came and joined in the fun. There was plenty to eat and drink, and I got a lot of referral names from the area. My supervisor would then take me out to "cold turkey" sell to these people. I learned, after about four months, that the art of selling was a lot more than just amusing the customers with the demonstration. The art was to close the deal!

Ora Young, the Bureau of Air Commerce inspector in the Chicago-Milwaukee area, 1937, and the author's mentor in his aviation career.

Jim Braden was the supervisor, and he was an artist. He could tell when to pull the order blank or pack up and leave. Braden took pity on me and put me in the Third Street Schuster store where Jean worked. This was fun. I was sort of a circus huckster — the more noise you could make to get people to crowd around, the better. The trick was to get their name and address to get a free demonstration, and they might win a hand-held vacuum.

All the time I was working for the Eureka Vacuum Cleaner Company, I was looking for work that suited my training and objective, which was the airport operation and more flying experience. I was fortunate to make contact with Al Tank, the president of Milwaukee Parts Company. He asked me what I could do for his company, and my answer was a brief review of my experience. I had a rough idea of what he needed to solve his problems.

"I know about flying, high-altitude weather analysis, mechanical engineering, general aerodynamics, propeller design, and performance testing in flight."

He seemed pleased and took me into the plant to show me an all-new four-cylinder inverted engine that was about to be tested. He wanted a four-bladed test club prop to move air and to load the engine. All I needed to know was the horsepower and rpm: about a 70 hp engine with a maximum

2,300 rpm. Mr. Tank made a deal. He would buy the wood blank — the pattern — to my design and agreed to pay me $150 if the prop did the job. This was my first break.

The wood shop that built the blank was the old Hamilton Standard Propeller Company that had been around for years and was expert in duplicating fancy wood statues and wood carvings for buildings and other applications. The company still had the old propeller carving machines. It took about ten days to design and finish the first club. I carved it myself. It was a standard shape and was finished like a piano.

Mr. Tank was surprised and pleased. He kidded me about it: "What if it over-revs?"

My answer was, "You have a fine engine that's putting out more power than you think."

"Oh, I see you've built a little slack into the design."

"That's true," I said. "The trouble with a test club is if it turns too slow, how much do you cut off each tip? Just for fun, I'll mark how much before we start. Do I have any gamblers in the group?"

Mr. Tank's remark was a classic. "Either you know how much power the little engine will put out before we've made a test run, or you believe in the supernatural."

There were three or four men getting the engine ready to start, rigging the fuel lines, air channels, etc. Mr. Tank asked me to come downstairs with him to his office and drawing room. He showed me some of the other engine designs, all V-8s, taken from the basic block of the World War I Curtiss OX-5. The idea was to convert the hundreds of engines that were installed on various airplanes; they were wearing out and no replacement parts were available. Mr. Tank's plan was to modernize those engines — improve their power and reliability — moreover, for one-third the cost of any new engines that were on the drawing boards or were in production in the 100 to 150 hp class.

The little engine was ready to start. In the late afternoon, I suggested that we slow-time it for a "break-in" before we made the final decision on how much to trim off each blade for a maximum power run. Mr. Tank agreed, and I promised to be back the next day. I did warn the mechanics to watch out when hand-propping the four-bladed propeller. "The blades are very close together and could nick an arm if you're careless."

The next day, I returned to the plant on South Sixth Street, anticipating that the engine would be ready for a maximum power run. Mr. Tank told me the sad story. The cast steel cylinder sleeves that were designed to be removable, but were pressed in the cast aluminum cylinder, were dropping down when the engine got hot. This meant a redesign of the sleeve and cylinder casting. This problem almost caused the end of the project. However, I suggested that new sleeves be machined with a small step at the base with proper clearance machined in the base of the cylinder casting. Mr. Tank agreed and gave the order to make the change. The propeller was on hold; and so was my $150.

Mr. Tank was ready to flight test for federal inspectors the Model 63 V-8 115 hp engine, but he had no propeller to match the engine and his Waco 10 airplane. The feds insisted on 100 hours of flight testing with the proposed propeller to prove both engine and propeller marriage. The engine had been government-certified after 150 hours test-cell running at the University of Minnesota. Mr. Tank asked if I would consider designing the master propeller; we could prove the combination on the 100-hour flight service test and use the same prop to produce copies. I agreed to do the design and make the templates for $250.

I found Iky Levine, a retired carver for the Hamilton Standard Propeller Company. He would do a master and some finish carvings and balancing on a few production props, and I knew his prices would be fair. I completed the drawings in two weeks, and Iky was finished in about one. The prop was perfect. Al Tank's test pilot, Ray Zulkie, was very pleased. I insisted that he run a series of airspeed calibrations, climb tests, and takeoff tests corrected to sea level. I privately told Al Tank that, in my opinion, a test pilot's notes should not be used until proven by actual data recorded independently of the pilot — either by a photo panel or by an observer aboard specifically for that purpose. I suggested that an engineering student from Marquette University might be the answer. He could ride in the front cockpit and take data according to a flight plan that the pilot must follow. We agreed that the airplane engine and propeller performance data were more important than the engine performance alone. And as long as we had to fly the combination for 150 hours, we should make full use of the time and expense. This would also make a good sales pitch for the engine and propeller later. In addition, I made it clear that I needed the true data to design other propellers for similar airplanes; it would cut down the time by 50 percent. Most of the OX-5 powered airplanes had similar performance, so blade-angle propeller changes would be the only element necessary in order to change the design to fit several airplanes like the Travel Air 2000, the Curtiss Robin, the Bird, the Waco 9 and 10, and the American Eagle. There were about 2,000 airplanes that would be a sales target.

The new four-cylinder Tank "Sky Motor," as it was called, was ready to test. After about four weeks, the initial runs were routine. However, it appeared that the engine would be hard to cool, though Mr. Tank was getting his 70 hp at 2,300 rpm, as planned; the club prop proved this.

We trimmed about one inch off each blade, and my comment to Mr. Tank was, "You guessed it right."

He looked at me with his typical smile: "A miracle that the engine is developing the 70 hp and a good engineering job on the prop." He continued, "I know we're going to have trouble cooling the engine, so I think I want to install it on an airplane that a friend has built." The aircraft had been designed for a Model A Ford conversion, and the engine was very heavy, requiring nose ballast. The new Sky Motor engine would reduce the aircraft's overall weight with ballast by about 100 pounds. The plane was a well-built small biplane and would make a fine test bed; however, we would need a new propeller.

"Would you consider designing a propeller for the combination?" Mr. Tank was a practical man, and he was thinking ahead of his accounts. He owed Iky Levine $150 for the Model 63 prop and $400 to me for the four-blade and the design of the Model 63 — all together $550 — and I felt responsible to Iky. I confronted Mr. Tank with the problem, and he tried to reassure me that he could clear his debts to me and to Iky within 10 days.

I wasn't certain this could be counted on, nor, in the long run, could a future, so I attempted to find a job elsewhere, keeping in touch with Mr. Tank to help where I could.

I was fortunate to secure a job with the Schuster's Department Stores as a supply floor manager, and although it was a long way from flying, I must admit, it was good experience and taught me about real people. My job was to settle complaints if the lady at the complaint department window couldn't. Polish and German housewives were a challenge. But I used the sweet-talk method, and they always went away smiling and thanking me for my help. The job gave me some steady money and a chance to think about getting married.

I was aware that the Milwaukee airport lease was going to terminate. The current leasee, I found out, was marginal, so I contacted the Airport Division of Curtiss-Wright in New Jersey. Their answer was favorable, but they would require financial and experience information to consider a long-term lease. The present contract was in force until 1 June 1935, which gave me about eight months to put my program together. I was determined to get the money, the Bureau of Air Commerce support, and the lease to try a fixed-base aircraft service operation.

The facilities were in good repair and were designed with ample office space for weather, engineering, repair maintenance, ground school, and flight training. A charter service was also possible because of the fishing and hunting in northern Wisconsin and Minnesota.

I made several contacts through the YMCA manager and board of directors, informing them of my plan. I also briefed my father and Al Tank and contacted Ora Young in the Bureau of Air Commerce office in Chicago. He simply remarked, "When you're ready to take over the airport, I'll fly in and look at your plan. Then we'll recommend what's required." All was encouraging so far.

I continued to work at the store, and business in general was looking up. Jean and I made the decision to get married after 1 January 1935 and not to tell our folks until we were settled into an apartment. We found a nice, but small, place for $35 a month, comfortably furnished. It was in a good neighborhood, close to work. We believed in the old theory that two can live as cheaply as one. This theory works pretty well until your wife gets pregnant. I had moved out of the YMCA on 1 January. Jean had stayed with a girlfriend from the store, and we were married 11 January 1935 in Waukegan, Illinois, Jack Benny's town — the "marryingest town" in the country. It was on a Friday, our only day off besides Sunday. It was also President Roosevelt's NRA (National Recovery Administration) day, the start of the five-day week in those Depression recovery days of the thirties. We made the trip to Waukegan on the commuter train and were back for supper at our favorite restaurant across from the YMCA. We stopped at a liquor store on the way home for a bottle of blackberry wine, drank the whole quart, and made it to work the next morning. We were happy, but realized the long, hard, and rough road that lay ahead.

Married Life

Married life agreed with both of us. Jean was a stabilizer for me. She frankly told me to get back to my flying, even if we had to leave Milwaukee. "If the airport deal is what you want, let's go after it." The next thing was a car. She made it clear to me that she had saved a little money and told me to go find a used Model A Ford Coupe. I had some free time from the store because I worked on a "call" basis. After trying all the used car lots, I finally found a 1932 Ford Model A with special balloon tires. The deal was $150 total. Down payment was $80 and with interest our payments were $7.23 per month for one year. Our insurance added $1.00 per month, making a grand total of $8.23.

The little Ford was the best thing that had happened to us since we were married. We drove around to get acquainted with the town — the well-kept parks, the lake front, and the beautiful residential areas on the north side called Fox Point. Of course, we spent time at the airport looking over the facilities.

To our surprise, there was an apartment on the north wing of the hangar with a separate entrance. The operator, Mike Bronson, was an old-timer in the flying business, but it was obvious that he wasn't much on organization. He was nice to us, however, and told us the pros and cons of the facility. Heating was the big problem in the hard winters. The large, separate heating plant was in bad shape and Curtiss was not interested in making repairs. He did say that the airport could be a paying operation if it were run properly. He also said that he was retiring and moving south to organize air shows across the southern half of the United States. Bronson was a typical old-timer. He loved to fly his Waco Taperwing every Sunday to do his own air shows, and sometimes he would have a local parachute jumper put on a death-defying free-fall. This would bring out a few people to take rides. During the week, in good weather, he would fly hunters and fishermen to some of the northern Wisconsin lakes and woods. He was a one-man show.

Jean and I decided it would be wise to plan a trip back to our homes in western New York as soon as we could get time off. The long Labor Day 1935 weekend, plus a week, would make a nice trip for us and would help our folks' frame of mind about our getting married "on the sly." I could also check on the money from my insurance and possibly borrow some from my dad. This would give me a basis for our airport plans.

We worked hard all summer and managed to save some money, enough to make the trip home. It seemed that our lives were becoming squared away. I worked on the car — ground the valves, cleaned the carbon, swapped tires, and aligned the front. We were ready to go by Saturday, 31 August 1935. We managed the time off from Schuster's, and I put off Al Tank on another propeller for the little four-cylinder engine.

Our trip was non-stop, except for gas and food. We arrived at Jean's folks in Honeoye Falls, New York, 30 hours later, about 3:00 a.m. Labor Day. We were tired but relaxed. We showered and slept until noon, then were ready to go see my folks. Jean's father and mother drove their car. The trip over the hills to Bristol Valley was beautiful. The leaves were beginning to turn on the maples and oaks. The grapes were almost ripe. Many fond memories lay in those beautiful hills for both Jean and me. The gliders and the big kite seemed a long time ago.

From the back seat, I overheard Jean and her mother talking about a lawn party and wedding reception for friends and relatives. I didn't respond, but privately thought the plan was a good idea.

When we arrived home, my mother was there but Dad was at the store in Bristol Center, about one mile south. I left Mother in shock; Jean and her folks stayed with her. I went to the Center to see Dad, Frank Tones, and Jim Reed, the store owner.

I drove up and saw that all the locals were on the front porch of the store. I yelled, "Where the hell is the poker game?" My dad came to the car and grinned and said, "You thought you could get away with marrying that nice girl and not telling your old friends."

It was a pleasant experience to be back home. I got the feeling that they all were glad to see their home town boy well and grown up. We finally left the store, and by the time we arrived home, the garden party was planned and the guest list was completed. Mother was to call the friends and relatives on the Rowley side; Jean's mother was to take care of the Austin side. My dad and Jean's dad were to take care of the food, such as hamburgers, and "white" and "red" hot dogs. Jean's dad insisted he would take care of the punch bowls and the corn on the cob. He thought it would be nice to have a tent in the backyard in case it rained.

I thought to myself, "What the hell is going on?" But I told myself, "This is the first time you have been married, boy!"

The party was on Friday, 5 September. Ralph, Jean's father, had borrowed a burial tent from the local mortician. This was a subtle joke, typical of Ralph. I wasn't aware that the two punch bowls were spiked — it was "sly" stuff. I noticed that all the teetotalers were laughing a lot. This included my mother, who had never touched a drop in her life. Ralph confessed later to the deed. But it was a beautiful party, and everyone stayed till late evening. Jean and I had to leave on Saturday morning to make work on Monday, 48 hours later.

We left about 7:00 a.m., Saturday, after sad goodbyes. The weather was perfect, cool and clear. Jean fell asleep soon after we hit the highway. Our first pit stop was the east side of Erie, Pennsylvania. We were averaging about 50 mph.

Jean didn't want to drive. She complained of feeling sick to her stomach — probably car sick. I gave her a couple of Alka Seltzers, which helped and she went back to sleep. But this wasn't like her. The thought flashed across my mind: "Wonder if she's pregnant?" This would change our plans, but good.

Jean came to when we got into Cleveland traffic. She was alert and relaxed when we stopped for supper in Elyria, Ohio. We were ahead of schedule by about three hours, so I slept about an hour before hitting the road. We wanted to make the Chicago area before the Sunday morning traffic. The trip on to Milwaukee was uneventful, and we arrived at the apartment on Sunday evening in time for a good night's sleep.

Monday morning we were up early and had our breakfast. Jean was sick again. Then she confessed that she and her mother had decided she was pregnant. I was in shock. All day I wondered how we could make ends meet on one paycheck. The airport deal was out. She would have to resign her good job. But there had to be a solution aside from going home.

We tossed the problem around for days. The doctor confirmed the inevitable, and I told Al Tank about my problem. He congratulated me and made a joke. "You're like all the young guys who dream of being a father and when you find out, you turn white and start to wonder 'why me' at a time like this."

I could see he was concerned, however. He changed the subject. "We've got to finish the flyable prop for the little four-cylinder. We have the engine installation about finished in the biplane test bed."

I reported that the drawings were completed and that I could finish the templates for Iky in three days. Al remarked as I was leaving, "Will you be available Monday morning?" This was Thursday. I would be available. "I want you to meet a friend, John Claude. He might have some answers to your problem. That is, if you don't mind working as a strike breaker in a stove factory."

My answer was, "At this point, I don't care, as long as it pays well."

Mr. Claude met me at Al Tank's office on Monday about 1:00 p.m. He wanted to go over my experience and education. The interview was brief. Claude wrote a note on the back of his business card to Jim Linderman, Vice President in charge of manufacturing at Linderman-Hoverson (L&H). I was to report to him as soon as possible.

I wasted no time the next day getting an appointment. I ran the picket line of south Milwaukee Poles. They were an angry, nasty group, but didn't touch me as I walked in the front door. If words could kill, however, I would have been dead.

The interview was brief. They wanted a model maker and designer in the electrical appliance division. The operation was under A. J. Rutenburr, a fine engineer. He talked to me about the model year changes as though I knew what it was all about. We toured the experimental department and the shop. He was obviously anxious to get started. He flat out asked me when I could come to work. My question was how much will the job pay and for how long? His offer was very fair. Forty dollars per week with overtime provisions. Overtime was going to be a must until the new models would be ready for the Chicago show.

Mr. Rutenburr had no question regarding my ability, so I had no reason to refuse his offer. I promised to come on the job in one week, but requested a parking space inside the plant. I felt the strikers would damage my car. This was acceptable to him, and I signed up. On the way home, I wondered just how lucky we were: pregnant, a new job, more money than I had ever made. The only negative thing was my flying future. The airport deal was off, but I was going to have enough money to do a little flying to keep my hand in and to get my commercial pilot and mechanic licenses.

Jean was pleased with my progress and planned our budget to provide for the expense of the baby. We decided to find less expensive living quarters as soon as possible. She was allowed to work three months before having to leave her job. We found a nice single room efficiency in a private home on the west side of town — a very pleasant change. Mr. and Mrs. Keohn, the owners, were like father and mother to us; they became lifelong friends.

The new job was a challenge. Mr. Rutenburr cut me a "bit of slack" for about a week, and then the heat was on: seven days a week, nine hours a day.

He was pleased with our progress. There was no slowing on my part. The older union men would work just so hard; therefore, my progress looked good by comparison. We completed the new models and tested them in three months. Mr. Linderman was very complimentary and made an effort to come to the lab and offer me a permanent job in the department. My reply was negative, however, because of my flying career. I stayed on with L&H until little John was born on 3 April 1936. The winter of 1935-1936 was one of the worst in history. There had been as much as four feet of snow in one storm, and to drive to work was a real challenge — usually two hours to make three miles. Work started at 7:00 a.m., and I didn't miss a day. I was fortunate that I had a lot of overtime, because as it turned out, the baby expenses ran more than twice our estimate. But we were pleased with him; he was a big, strong kid.

Spring came early that year. By the end of May, life in Milwaukee was back to normal. The snow was gone and the flying weather was good. I had taken advantage of the long winter and for $10 an hour had rented an OX-5 short-nosed Eagle from Ray Zulkie, a private operator. Five hours put me back in practice and ready to get my commercial pilot's license. While I was flying around the Curtiss-Wright field, the new manager, Elmer Frankie, contacted me. His proposition was a 50-50 partnership for $5,000. I was to be the active manager and live in the apartment. We would split the profit evenly. His wife would keep the books. He agreed to work weekends and instruct students, and I was to do the mechanical work on airplanes stored in the hangar as well as handle transits, fuel, and oil sales. This deal sounded good but turned out to be a real strain. I did gain a lot of experience, which was a plus in my future flying career, but I also learned never to be a 50-50 partner in any business venture.

19

Milwaukee Airways

Elmer Frankie, my partner in the airport venture, was a good pilot but was a typical city employee in Milwaukee's water department. He was on call 24 hours, and there seemed to be a rash of water main leaks the day I went to work as airport manager. My work hours were from daylight to well after dark.

Jack Stein, a young man who was working on his mechanic's license experience requirements, proved to be a godsend. He was talented and didn't watch the clock. He took some of the detail workload of my daily schedule, giving me more time for flying and for studying for my commercial pilot and mechanic license tests. I was determined to pass both written and flight exams with top grades. There was some pressure from Elmer to get my mechanic's license so that I could sign off mechanical work on aircraft and engine repairs. We also needed a repair station number. This would pull in some outside business and would give Milwaukee Airways a national listing on government records.

I was fortunate. I met Ben White and Hans Krimsreiter, partners on a very clean Beech Travel Air 2000 with a Tank model 63 engine. The plane was based at our airport, but they flew it very little. Ben was the pilot, and Hans had the money and was full of ideas. Ben was also a good mechanic. They both came to me privately to discuss a patent idea on an expanding airfoil. Al Tank had recommended that they contact me to do the development engineering and wind-tunnel testing to verify patent feasibility. Their proposal was to credit me with one hour of flying time for every three hours of engineering time. I agreed if the outside expenses for wind-tunnel time and model fabrication would be absorbed by them. I ran this past my partner. He

The Greve class racer, designed by the author in 1936 for private buyers Ben White and Hans Krimsreiter.

agreed that this would give us another airplane to use on the weekend passenger rides and short hauls around the county.

I started fabrication of a 24-inch, 50-mph wind tunnel and a suitable balance unit to measure lift and drag in grams generated from a 12- by 2-inch model wing with different thicknesses. I selected a well-known Clark Y basic airfoil and added 5 percent thickness to each of three model wings, making the third model 15 percent thicker than the original. The leading edge radius was the same for all. Ben and Hans became so interested in the project that all I had to do was make the drawings. I did have to make the English-type vertical balance device at Al Tank's shop, however, and even Al got into the act. We worked night and day, and in less than two weeks we had the aerodynamic proof that Hans and Ben were right. The idea was feasible, provided we could design an actual wing section that was expandable in thickness while in flight.

Ben called a meeting of the patent lawyer, Hans, and me. By this time, our inventors were trying to put the "lid on" as they were aware they had a good idea. I assured them they were safe as long as we didn't expose the data and the wing design. The wing was a problem until I realized that we could build an expandable upper camber rib by using a flexible steel cap strip with a series of scissorlike levers connected to a long link. The link was moved fore and aft by a torque tube that could be operated from the cockpit. I made a sample rib for Ben. He fabricated an 8- by 3-foot all-metal wing stub that worked perfectly.

My pay was 60 hours of flying time in the Travel Air. The patent was filed and granted within three months, which must have been some sort of record.

Ben and Hans had another project that they wanted me to engineer: a racing airplane in the Cleveland air race Greve class. The engineering had been completed except for the engine installation, cowling, and propeller. This was not a high-priority project; I could complete it in my spare time.

After about 45 days, I was sure of myself to the point that I contacted Ora Young for an appointment to take the tests for my government certificates. The moment of truth was 2 August 1937. I guess my apprehension was high because Young had a reputation of being a no-nonsense, critical examiner. I was relieved when we sat down to review my logs and experience. He let me talk; he didn't want any embellishment, just the facts. He apparently was pleased with my response, and told me, "Let's get trucking. You have a hard day ahead."

He pulled out both written tests, pilot and mechanic, then shocked me by saying, "I think you can do both today. I have to be at the Municipal Airport for a flight test. I should be back after lunch. I'll see you then. This is not an open book test. Good luck."

The pilot's exam took about four hours; the mechanic's, four and one-half hours. Young returned at about 2:00 p.m. I had finished the pilot's written. He picked it up and remarked, "I'll grade it; you had better get some lunch."

When I reported back, he remarked, "You passed ninety plus. You better get on the mechanic's." It was tough — a lot of essay writing and detail. While I was working, Young inspected the shop and hangar area for the repair station certificate.

I completed the mechanic's portion of the test at about 5:30. He remarked that he would read it later: "Will you be ready to take the flight test tomorrow?"

"You set the time and the place, Sir," I said.

"How about Municipal Airport at 8 o'clock." He left. I reviewed the day and decided to fire up the Travel Air and do some fancy flying. After about 30 minutes, I was relaxed and decided to come in and "chow down" and call Jean, who was on vacation in New York. I made a full report on my progress to date, in particular that I thought the written was over and tomorrow was the final grade review and flight test. The family wanted to know when I was coming home to stay. I had heard this before and for some reason it was starting to make sense. If I passed the flight test, the mechanic's written, and the certified engine and airplane repair station, I would have accomplished my objective.

The final day dawned cool and clear. No clouds. I checked out the airplane very carefully. I called the tower at Municipal and got an approval for a low pass over the field at 8:00 a.m. I figured Mr. Young would be in the tower watching for my landing, and if there was no traffic, a slow roll would be in order. It would add a little spice to the day. I had done this for my old instructor, Otto Enderton.

Conditions were perfect over the field. I chose 1,000 feet above the ground. I saw the green light: a left roll a little downhill felt good, speed perfect, 180 degrees overhead approach; the landing was on the spot. As I taxied up, Young was standing on the ramp in front of the tower with his helmet and goggles on. He motioned to my spot and to cut the engine.

He gave me a briefing on what maneuvers he wanted on the test. I wrote them down in order on my knee pad which impressed him. He remarked, "That looks like an Air Corps note pad."

"Yes, Sir, I used it flying the weather airplane at Scott Field."

"We'll talk later," he said. "Let's get flying."

The flight test was routine until the spin test. The left spin was uneventful. I climbed back to 3,500 feet and pulled up in a stall. Just prior to the right spin, a loud click-click came from the engine. Mr. Young looked back and pointed to the engine. I nodded my head "yes." I throttled back and completed the spin. I figured we could work on the engine on the way back to the field. The engine would pull enough power to keep us flying to the landing pattern. I used the 180-degree overhead emergency approach pattern. The engine idled okay.

We taxied back to the parking area and I could tell the noise was coming from the left bank. I shut the engine down. Mr. Young deplaned and walked up to the rear cockpit with a grin. "What the hell was the noise?"

"Valve action," I said. "I hope nothing expensive."

We went to the Civil Aeronautics Administration (CAA) office to review the test. Young was very complimentary and assured me I would have my "tickets" in a few days. Then we spent the next half hour talking about Otto Enderton. He wanted to know how Enderton was, did I keep in touch, and added, "Your flying is a dead replica of his with a touch of show-off. I saw the slow roll when you came in. I also liked the way you handled the engine problem. No panic — that's good."

Mr. Young was in a hurry to get back to Chicago. As he left, he handed me a picture of himself standing in front of his Stinson. He also shook my hand and remarked, "You have a good flying and mechanical sense. Use it well."

I went back to my engine problem. To my great relief, it was a loose rocker arm adjustment, and I had the tools with me to fix it. We were ready to fire up and go back in about 15 minutes.

Elmer Frankie, my partner, was there when I returned. He was pleased to hear that we were a certificated repair station and that I had passed my pilot and mechanic licenses.

But a remark that Elmer made nagged at me: "Now that you have your licenses, we can expect more profit. You can do some of the flying and sign off on mechanical work." My attitude was purely negative, I guess, because I could see very little future. I was sure, however, that my career had gone one great step forward.

I called home to check on how Jean and the family felt about selling my interest and moving back East. The answer was, "Sell, but try to make some money for your trouble." When I approached Elmer, he seemed relieved. It was clear he didn't like my pushy ways, and he wanted a partner who could afford this expensive playhouse. He frankly said he was more interested in holding the lease for the control of the land. With a wealthy partner, he could purchase or get a long-term lease. He asked me to set my price and I did. Eight thousand. Cash. Clear of all bills and any future liens, etc.

To my surprise, Elmer had a buyer in less than a week — Bob Church. He didn't question the amount. The only concern was about Jack Stein, our mechanic. They wanted him to stay on a fixed salary. Jack was willing if I would verify his mechanical experience in writing. This was done and I called Ora Young to give Jack my blessing.

The sale netted about $2,000 which was not bad for a few months' work. The bonus was the pilot and mechanic licenses. Next to my military training as a meteorologist, they were most important to my future career.

I packed up our belongings, shipped some, and carried the personal items in the car. Saying goodbye made me realize how many friends we had. Al Tank, Mr. and Mrs. Keohn, Ben White, and Hans Krimsreiter headed the list. Ora Young also had become a friend. I ran into Ora in St. Petersburg, Florida, years later in the early seventies, at a Quiet Birdman meeting. We had a ball talking over old times. He was retired from active flying. I kidded him about having the reputation of being a tough examiner. However, he had the distinction of giving me my pilot and mechanic licenses in the same day.

I arrived back in Honeoye Falls on 10 August 1937. Jean and little John were as glad to see me as I was them. It was a happy get-together with Jean's father and mother, as well as my own mother and dad — a happy time for us all. The fact that I didn't have a job didn't worry me; I was glad for the vacation.

Curtiss-Buffalo

The Finger Lakes part of western New York in late September and early October is truly a heaven on earth: beautiful tree colors, the fragrance of grape pressings, and clean, cool air. The hills and lakes are well named the "Switzerland of America."

Jean, little John, and I spent 45 days with our folks and friends, in particular showing off little John. It was, however, getting obvious to me that our vacation was over and I had to get moving on my aviation career. I had been renting a Taylor Cub every weekend to barnstorm down in my home valley, taking friends and paying customers for rides over the beautiful hill country. But Jean reminded me that it wasn't a profit-making operation, and that our funds were getting low. I was feeling the pressure to get moving and thought it wise to contact Otto Enderton, my first flight instructor. He was then the senior Department of Commerce flight examiner for Curtiss, Consolidated, Bell, and fixed-base operators in the western New York area and would know the jobs available in the local industry.

Otto was pleased to see one of his students. It had been almost two years since we had visited, so there were many questions to answer. He wanted to know what I was doing in Rochester and if I was flying for a living. The conversation gave me an opportunity to lay my problem on the table. I frankly told him that I wanted to move back to western New York and stay in the aviation industry. Our young son and the grandparents had a strong bearing on this decision. Otto was quick to point out that flying for a living in the area was out. All the test pilots and airline pilots were well entrenched. There were no openings. He went on to say, "I feel that you could benefit by engineering, manufacturing, and experimental experience. This would help in future test pilot job applications." He added, "In general, most pilots, even in high positions, don't know what makes an airplane tick. This situation has got to change. I'm doing what I can by requiring more engine, airplane, and engineering in our pilot's written examinations."

Otto thought that one of the Buffalo companies would have a spot for me. Curtiss, he felt, was probably the best. He suggested that I contact Frank Gifford at the Curtiss hangar at Buffalo Municipal Airport, in the hopes of meeting Lloyd Child, the current chief pilot. Don Berlin was the chief engineer, Pete Jensen was the plant manager, and Bill Davey was the plant superintendent at the time. Jack Davey, Bill's brother, was head of experiment and development, and Stanley Vaughn was one of the old Glenn Curtiss employees who ran the actual experimental department. All new aircraft were built in his shop, Otto told me.

Otto also mentioned that he had heard that there were three new models in the works, one top secret. "I envy your age and the new developments ahead. It's the right time, and this may be the best place to start. Keep your flying current. It's like the violin; keep practicing. You'll need it one day."

Otto's words were music to my ears. He was so right, as it turned out later.

It was Monday, after my meeting with Otto, that I announced at breakfast to Jean and her folks that I was leaving and would be back when I had a job. This was over-dramatic, but it got everyone's attention. I did explain to Jean later that I was going to Buffalo to check on a job at Curtiss. I was on my way by 10:00 a.m. This was a thrill because I had a feeling that this was another big step.

I stopped at the Woodward Airport in LeRoy, New York, to look around and kill a little time so that I wouldn't catch Frank Gifford out to lunch. I had no appointment. It would be pure luck if I got to see him the same day. I arrived at the Buffalo airport about 12:30, had lunch, and cased the area, in particular the Curtiss hangar. The weather was warm, so the hangar doors were open enough to see inside. In plain view was a beautiful, all-metal, twin-engine attack class Air Corps bird, a YA18. I decided that 2:00 would be a good time to put in an appearance.

There were no guards, no security. I walked in the hangar and asked the first man I saw, "Where can I find Mr. Gifford?"

He pointed. "He's in his office, over there." Gifford's door was open.

As I approached, he looked up from his desk, smiled, and said, "What can I do for you, young man?"

I introduced myself and added, "Otto Enderton recommended that I see you, Sir. I'm looking for a job."

"This may be your lucky day," he responded. "Jack Davey is looking for an engineering project coordinator. What can you do?"

"I fly, have a commercial license, and over 1,000 hours certified," I said. "In addition, I'm an A & E (Airplane and Engine) Mechanic. My education, a degree in meteorology from Army Tech School, Fort Monmouth, New Jersey. I was transferred from school to Scott Field, Illinois, and was assigned to the upper-air recon aerometeorograph development group. I also have had two years' fixed-base experience and was part owner and manager of Milwaukee Airways, operating from the old Curtiss-Wright airport in Milwaukee. Otto Enderton taught me to fly in 1929. I also graduated from the Times Union School of Aviation in Rochester. From there, I went into the Army in 1931. While I was stationed at Scott Field, I attended Washington University's engineering extension courses, math and science. No degree."

Mr. Gifford called Jack Davey at the plant in Kenmore, New York, and flatly said, "I think I have a man for the coordinator's job." Mr. Davey apparently said to send me over, so I got directions and was on my way. Mr. Gifford became a close friend and, in many cases, my close adviser during my stay at Curtiss.

Jack Davey was an easy man to talk to. He asked for a few details and handed me an application blank. The final interview took about an hour. He offered me the job as engineering coordinator on the SBC-4 modification program, with a promise of the "Turkey Hawk" project — a modification of the old CW-22 — that was to be transferred from the St. Louis plant. It was agreed that I start work on Monday. I left the plant and headed for Honeoye Falls with a job and good pay — all in one day.

I was the 401st employee of the Curtiss-Wright plant in Buffalo. This was announced, and my name was given on the news bulletin that was broadcast over the factory's PA system every Friday. During the Friday bulletin our problems, failures, and successes were announced. I thought that this was a fine idea, to keep the employees informed; it gave everyone a sense of being on the team. There was no union in the Curtiss-Wright plant and never had been. There was no labor problem that wasn't solved on the spot.

The folks on both sides were pleased that they could see little John often and have us all on weekends. I was relieved to be close to home because my work was going to be a challenge with long hours, at least for the first few weeks, and the folks could help Jean find housing. They rented an "up-and-down" flat near some of Jean's relatives. When the real winter came, we found that you should never rent the lower flat — heat goes up. We were heating the top flat as well as our own. But after winters in Milwaukee, we didn't seem to mind.

From the beginning, Jack Davey was a friend and a good boss. He saw to it that I knew every person who was important. He spent time with each, explaining what my position was to be, both from an engineering and a manufacturing viewpoint. As it turned out, this was good for me. It left no doubt who was in charge of the modification of the SBC-3 to the SBC-4 configuration, and I had no trouble meeting schedules through the whole program, including the flight test, though it occasionally took a little "butt-chewing" on my part, which got me the name of "Cactus."

Jack Davey helped me in some of our project meetings with Pete Jensen, the general manager. Pete was fair but super-rough. If you didn't have the solution or failed to give him a straight answer to a problem, you could be in trouble.

I remember one particular incident. The new SBC-4 pilot's instrument panel had been rearranged with additional instruments so that with the original shock mounts, the panel would not pass the vibration tests. This problem was holding up the first flight schedule. Engineering was slow getting the change order out; their problem was too much theory and not enough practical sense. So I took it on myself to make "bootleg" changes over the weekend. The fix cured the vibration problem.

On Monday I told Jack Davey. His very pointed remark was, "Did you get the Navy inspector to approve the fix?" Of course my answer was "no." Jack remarked, "I can't help. You have committed one of the great Navy sins. Installation without inspection. I suggest that you confess to Mack MacKine, your Navy line chief."

I didn't expect this reaction. I went first to Pete Jensen and privately told him the story. He put on the same show that Jack did. I was getting the feeling that Jack and Pete were trying to teach me a lesson. My response was, "I'll take the fix off and take it to engineering and wait for the drawing and retesting. I guarantee this will cost us at least one week. If we can get the fix approved for first flight only, we can take the bird to the airport and our initial flight can be on schedule."

I started to leave the office. Pete called me back. "Sit down and listen. Young man, you're new here so I can excuse a basic error in Navy rules. However, you have put your finger on a problem that bothers our progress on experimental Navy projects. Tell Jack that we will call a meeting with the Navy's senior representative. We may get your first flight idea approved. Meanwhile, you tell engineering about your idea; make them think it's theirs. Get them to make a drawing of the proposed fix. I repeat — the *proposed* fix! Do you read me?"

My next move was engineering; Ray Blaylock cooperated. We had a proposal drawing the same day. Mack MacKine, our Navy line inspector asked me what was going on, and my reply was, "Mack, I think we have a fix for the instrument panel. Blaylock has a drawing. I'll put the parts through inspection as soon as possible." That night, I pulled parts off the airplane after Mack had left. The next morning, I took the parts to Blaylock. We checked them to the drawing to be sure it was correct. Then I put a work order in process to build new parts.

I stopped off to tell Pete what had happened and to my surprise, Mack was there. He was mad as hell! He had checked the airplane the morning after I had removed the parts. I let him blow off steam. I explained that the fix was installed for test only and that the parts were sent to engineering after the test. The test was run over the weekend to gain some time. I could see that Pete was getting ticked off with Mack's attitude. The big Swede stood up and flatly told Mack that he would talk to the Commander about changing line inspectors if he was going to obstruct honest efforts to get the airplane back on schedule. It was time for me to play the poor young kid. I stood up for Mack and confessed that I had jumped the gun. Pete looked at me. I saw the glint in the big Swede's eyes. I had played the right card. Mack had his fun and I was off the hook. Pete, Jack, and I talked the caper over and allowed that we should be in the movies! Our meeting with the Navy representative was not necessary.

The SBC-4 modification program was getting close to the initial flight tests. I would be at the airport all during the first experimental tests. Mr. Gifford was anxious to get the flight phase under way and wanted to see the flight plans. The Navy had very definite regulations on flight-test procedures. The documents were in the Curtiss-Wright library, so I checked them out and worked up a flight-test outline, including instrumentation requirements. For some reason, the flight-test department did not include flight-test engineering personnel or instrumentation experts. Engineering was supposed to handle this work in their test lab. The test instrumentation was antiquated. Frank Gifford asked me to look into the matter because the AT-18 engine modification program had trouble getting good data and had suffered some flight delays; he didn't want this to happen on the SBC-4.

This problem was too hot for me, however, because it involved Chief Engineer Don Berlin, Chief Pilot Lloyd Child, and Plant Manager Pete Jensen. Jack Davey, as head of experiment and development, needed to be the man to set up plans to change the system.

I reviewed with Jack what Frank Gifford had told me and the flight plan outline that I had prepared for the SBC-4 flight tests. Jack was shocked that I had become involved in this level of company policy; it was unusual for an engineering coordinator to get into the flight-test plans. But Jack finally agreed to ask Pete Jensen to set up a meeting on the subject. And he asked me to write a memo to him outlining the problems and suggested corrections. The memo was a success, and Jack sold most of the ideas to our management. The important change was to move the instrumentation lab to the hangar, with Frank Gifford in charge of the personnel and the new state-of-the-art test equipment. Everyone agreed that this was a good change. Engineering also assigned Frank Lakowitz full time to flight test at the field.

The first SBC-4 test airplane was ready to fly on schedule on 15 February. Lloyd Child made the first flight. The temperature was 10 degrees and CAVU. Wind was about 8 to 10 knots on the runway. I had been at the airport since dawn, and Frank, the ground crews, and the mechanics were there by 7:00 a.m. Lloyd arrived about 8:00.

Lloyd Child was a tall, quiet man, the type you just can't figure. This time I was determined to get him to talk, so I walked into his office and asked if there was anything that he needed. He was suiting up. He looked up from his muklucks. "Yes, how was the run-up?"

"Everything checked out, flight-test radio, cockpit heater fair," I said. "I have a question, however, about the carburetor heat. Personally, I think it's marginal, but we have low humidity, which should help."

Lloyd thanked me for the briefing and walked to the bird. Frank Lakowitz and the crew were there. Lloyd asked, "Giff, will she fly?"

Mr. Gifford's response: "It better or my you-know-what will be the reddest." Lloyd grinned and climbed aboard.

Mr. Gifford asked me to go along in the airport pickup to the runway. As we drove, I asked him about Lloyd. He remarked, "He's a cool cat. Under that quiet, low profile, is a guy that knows his business and if something goes wrong, it better not be a stupid mistake. I've been in my job for years and have learned his ways. For instance, if he likes the bird on the first flight, he will make a fast pass down low and do a slow roll. That's his way of saying thank you to our ground crew and everybody involved. He also will fly out over Lake Ontario out of sight and climb up to at least ten thousand and put the airplane through its paces. He stays out about one and a half hours if everything is okay."

We could see the bird coming overhead in about an hour and 30 minutes. He was downhill to make a fast pass and, as predicted, he did a slow roll. We all were pleased. Lloyd had very few "squawks": hard brakes and low carburetor heat, but stalls and rigging okay. He wanted to talk to me about the carburetor heat.

"Would you follow this problem with Ray Blaylock? We need about 50 degrees Fahrenheit more heat rise. I'll have the flight report tomorrow." Lloyd added, "I hear you're a pilot. Would you like to take a ride?"

"Yes, Sir, if I can do something useful."

Lloyd replied, "Good. We'll do just that, after we get a little farther along with the program." He added, "You guys should be proud. This program looks good."

In about three weeks the program was well under way. My serious work was done, and I wondered what was next. I soon found out.

The Curtiss Turkey Hawk

The SBC-4 program had launched me into favor with the overall management of Curtiss. In addition to the name of "Cactus," I was called "Mr. Can Do." And even Pete Jensen, "Mr. Rough," knew who I was. The SBC-4 program was running smoothly. My overtime was finished, so weekends and evenings were free. I took advantage of this and spent time at the Consolidated Airport nearby. Dick Benson put me on as a part-time pilot for short hops, weekend passengers, and primary instructor. When the weather was bad, Jean and I would pack up and head for the folks' on Friday evening. This was a great time for us, but we knew it wouldn't last. The Turkey Hawk aircraft was on the horizon, and I was hoping to get the production coordinating program.

It was late Friday afternoon, around the first of June. I was paged: "Report to Mr. Jensen's office." It had to be something big. I knew it wasn't the SBC-4 program in general, but it could be the Hawk. I reported immediately. The first man I saw was Lieutenant Harvey Grey, a Reserve officer with whom I had flown at Scott Field when he was flying his four hours per month. The meeting consisted of Charles France, the general manager of the St. Louis Curtiss plant; Harvey Grey, chief pilot at St. Louis; and Pete Jensen, Jack Davey, and Lloyd Child of Curtiss-Buffalo. Ralph Kenyon, who was to be the production and tool planner, arrived late.

Mr. France gave a very thorough briefing on the aircraft, pictures, production breaks, tooling, and performance, etc. We were also apprised of the contract with the Turkish government. They were to furnish inspectors and pilots; the pilots would fly final acceptance. Our pilots were to make all initial production tests and would be responsible for the Turkish pilots' flight check. All tooling was to be transferred from St. Louis.

Mr. France asked Pete who was going to be the project coordinator. Pete said, "Charlie, we wanted to see the full scope of the job before we make that decision. We'll settle this before you leave Buffalo." The meeting broke up; Harvey Grey and Lloyd Child were outside Pete's office.

As I walked by, Harvey grabbed my arm and told Lloyd Child, "Lloyd, do you know about Elton? He's a fine young pilot. He flew the weather airplane for Scott Field's weather center for about two years. You should use him on the Turkey deal."

Lloyd replied, "He and I are going to have a go at the SBC-4, Harv. You may have a good idea." My private reaction was that I probably had no chance because of the other pilots on his staff. Lloyd made no further comment. Harvey and I went to lunch and talked about the Turkey Hawks.

"How did it fly, Harvey?"

"It's good," he said. "Fast and maneuverable. She's short and high on her legs — quick to ground loop. Be sure to be alert at touchdown. If you get the job, we can fly the prototype when you come to St. Louis. We'll have a ball. How about going back to Scott Field? You'll enjoy the plant. The CW-20, the pressurized twin-engine transport, has been revived. We also have a small twin-engine trainer on the board. You remember Dutch Kleinoder? He took a permanent commission and was transferred to the training command in San Antonio. Sergeant Finley is still in Operations."

Harvey then said he'd see me later. "Hope you get the job."

I went back to work; I was relaxed and confident that I had the position.

The next day Pete Jensen called the group together and announced that I was to be the Turkey Hawk project coordinator; Ralph Kenyon would be in charge of tooling and planning.

Ralph and I arrived in St. Louis at Lambert Field on 14 June 1938. We had about ten days to review the project and ship drawings, tools, and parts to the Buffalo plant. Kenyon was a good partner; he knew his side of the work, which was fortunate because the tooling was inadequate to the point that he had reported to his superior that new hydro blanking and forming dies would have to be made for the wings and aft fuselage parts. The old tooling was strictly hand-shaped and hand-formed, which caught my attention because of the labor cost and time schedule. My instructions from Jack Davey were, "Get everything shipped as soon as possible. We'll face the tooling problem later."

I spent some time at the Curtiss-St. Louis plant with Harvey Grey. The two-place Turkey Hawk prototype was flyable, so Harvey checked me out. The airplane was fast and would do all the aerobatics for a good fighter trainer. But its faults included poor stall characteristics and "tricky" landings because of rudder control loss at touchdown. The aircraft was high on its legs, as Harvey had said, which caused the rudder to fade out as the tail came down during landing. This problem showed up enough during stalls that I questioned Harvey about the advisability of putting the aircraft into spins. His reply arrested my concerns. In the drawings was a new, larger fin and rudder design. This new design, however, had never been flown. All this information was in the flight-test report documents. But I was concerned about the problem, so I called Lloyd Child and suggested that the prototype be flown to Buffalo to be modified and tested. Lloyd bought the idea. I was also concerned about the overall schedule. The Turkey Hawk was not going to be an easy project. This was obvious.

We had a weekend off in the middle of our stay in St. Louis, so Harvey and I went over to Scott Field for old-times' sake. Colonel John Peglo had retired and was living in California, Missouri. Captain Schoffield and Sergeant Farrell were in command of the weather center. Sergeant DePussen was flying the weather airplane, and Jim Crawford was working the aerometeorograph calibrations and data transmission. Schoffield, Farrell, and DePussen invited us to lunch; it was like old-home week as we reviewed each other's past four years. Farrell was married and was planning to transfer to the National Weather Center. Captain Schoffield was going to retire, as was Sergeant DePussen. There was little change to the base operations or buildings; the lighter-than-air hangar was still operating but at a reduced level. The grass field was still intact and was maintained like a golf course.

On the way back to St. Louis, Harvey asked me what I was planning for a career. My answer was, "Hang out my test pilot's shingle. In addition, improve my education."

He agreed, and also told me, "Your flying is very good. Why not ask Lloyd Child if you can do some of the flying on the Turkey Hawk? You could run the project as well. This experience would be good for your future resumé. You should plan on your final position or your career subject before you're thirty-five."

The next day, Sunday, we all went back to the plant to look over the Turkey Hawk prototype. Ralph and I had talked over the fin and rudder change; we were both concerned about the flight and structural implications — in particular, the aft fuselage just ahead of the stabilizer. At that point, the fuselage was very small in diameter — the same as the CW-22. We agreed that there should be a complete structural analysis on the rear portion of the fuselage before any parts fabrication, though this would cause at least a 45-day delay, even if the flight test proved out the change.

Our work in St. Louis was coming to an end, and I spent the last two days with Harvey Grey. One thing that stands out in my memory was Harvey's tour of the experimental building. The big show was the mock-up of the CW-20. It was going to be the largest transport airplane in the world. The cross-section of the fuselage was to be two circles imposed on each other forming a figure eight. The floor line of the fuselage was at the juncture of the two circles. The passenger version was to be pressurized. To me, the program was mind-boggling. While Harvey was showing me around, Mr. France and another man, Eddie Allen, came into the experimental area. France introduced me as the project manager of the Turkey Hawk. I had heard of Allen before; he was adviser on the CW-20 project and was to fly the first bird if it was built. He was very friendly. Harvey told him that I was a budding test pilot, that I had experience as a high-altitude weather recon pilot in the Army, and that hopefully I would fly the Turkey Hawk in Buffalo. At lunch, Mr. Allen talked about what was ahead for the new age of technical test pilots. He said, "We have some of the greatest stick and rudder pilots in the world, but few can analyze their own data after a flight." He advised me to take all the specialized training possible before I would hang my shingle out to the industry. Little did I realize how important Mr. Allen would be to my future flying career.

Ralph Kenyon and I finished our work and mailed our reports to our superiors in Buffalo. We arrived back at the plant early on Monday, 15 July 1938, reported in, and to my surprise, Jack Davey was in shock. He reamed me good about taking the liberty of reevaluating the flight testing, structural analysis, and tooling, as well as the resulting possible delays in the scheduling and the cost of the project. I listened carefully and retorted that he could have my resignation at once. This was my first real showdown with Jack. I had known for some time that there was one coming; this was it. The advantage was on my side. Jack took my resignation but asked that I take a few days to reconsider. He was trying to put me on the defensive. I refused, packed my belongings, and went around the plant to say goodbye.

Pete Jensen raised the roof when I told him. His reaction was, "What the hell makes you so hard to handle? The company has too much invested in the Hawk project and you to allow a stupid disagreement to affect the project schedule. I suggest you try to patch up your differences with Jack and stay put."

I then talked to Lloyd Child who assured me that he agreed with my report regarding the flight test of the prototype and the structural analysis of the aft fuselage. But he also added, "You know I make the decisions on all flight matters in this company!"

I went home that evening with mixed feelings and talked the problem over with Jean. Her remark was pure and simple. "You have been in the service. You should know the rules as to chain of command. I think you're operating like you were trying to find another Colonel John Peglo."

As I thought over the earlier SBC-4 caper with Mack MacKine, and now the Turkey Hawk modification, I came to the conclusion that I *should* slow down and follow the chain of command. I had been too busy and was not keeping in close touch with Jack Davey, my real boss. My dad would have said, "You got the big head."

I finally decided to talk with Jack and, as Pete Jensen suggested, to try to patch up our differences. I went in early and walked down through the plant to the SBC-4 line. They were just starting. Everyone asked where I had been, and I simply remarked, "Been doing some work for Mr. Davey out of town." I didn't say what, for a change. I went to the office about 8:00 a.m. Jack arrived about 8:30.

His first words were, "Glad to see you back. I hope you plan to stay aboard." Before I could answer, he said, "Why don't you take a long weekend and take your family down to the country? Come back Monday. This will give me a chance to go over your report along with Kenyon's. I have to coordinate the whole thing with all departments involved." Jack deliberately let me off the hook. I thanked him and went home.

After a lot of talk and thinking with both families, they all agreed that I should try to stay with Curtiss. This was good advice for the moment, but a long-range career there was out of the question. The company was old and the politics too strong for a newcomer to make progress in a reasonable time. For me, from then on, Curtiss-Wright was a stepping stone. For the moment, however, my job was the best that I could get for experience and training. I decided to finish the Turkey Hawk program and move to another position when appropriate.

Jack Davey and I discussed, in length, my position on the Hawk program. It was clear to me that Jack's problem with me was communication; I had been trying too hard to do my job with a minimum of contact. He asked me for more direct communication and a weekly written report for the first two months of the program. He also insisted that I spend more time with engineering and flight test on the prototype, particularly in the program's early stages. I knew from experience on the SBC that Jack was a little weak on the flight test and engineering; and he as much as told me so when he made his suggestion. This pleased me because this was where I needed the experience. In closing our meeting, Jack praised the report Ralph Kenyon and I had prepared on the Hawk program and assured me that he would proceed as it was written. His final suggestion was that I take a correspondence course on business administration either at the University of Buffalo or at the Alexander Hamilton Institute. His final comment was, "You'll receive a ten percent raise in your next pay envelope." We were paid in cash in those days. Jack made his point loud and clear. From then on we had a good and friendly relationship again. On his suggestion, I took the Alexander Hamilton course and learned a lot.

After careful coordination with the engineering and flight test sections, I set up a realistic schedule, though the first production airplane was going to be about 30 days late. The representative from Turkey was advised and was not pleased, but Jack handled him with a firm hand. "We'll do the best possible and will complete the contract on time. However, our expert, Mr. Rowley, says the first airplane will be late." Mr. Alonis Andari, in broken English, asked me, "Are you ever wrong?"

"No, Sir," I responded, "only occasionally, but not on this problem."

Jack told Mr. Andari to contact him at any time and added, "I understand that your pilots and engineers will have to be rescheduled. For any details, feel free to contact Mr. Rowley."

From that time on, Mr. Andari was with me on every move that I made on the program. He was a very interesting gentleman with a fine flying background in the Turkish Air Force. His retired rank was equivalent to a U.S. brigadier general. We spent a lot of time with Lloyd Child, particularly after the prototype arrived from St. Louis. The General asked Lloyd if he would check him out in the prototype, and, in his own quiet way, Lloyd explained carefully that the current configuration of the prototype would give the wrong impression of the final airplane, as there would be several engineering changes that would improve the stall and landing characteristics. Later, the General asked me what Mr. Child was saying. I reported the problem to Jack Davey who suggested that we arrange for an engineering briefing for the General and that I answer any questions he might have after the briefing. This was a good idea because General Andari's interpretation of our language was poor. I had been with him almost every day, however, and he seemed to understand either my English or my German very well.

The General's big problem actually started when his staff arrived: a male and a female pilot, a male and a female engineer, a male mechanic, and a male contract administrator. They spoke very little English as they were all directly from Turkey. We all reacted to the female pilot and engineer with wonderment — how were they going to fit into a pretty rough group of men with typical male banter? But we all adjusted and the women were highly skilled. My problem was to get the pilot checked out first and then he would decide how to check his pilot assistant. Lloyd Child would accept the plan only if the General used the initial production two-seater, so I arranged to have a two-seater produced first.

The Turkey Hawk program was falling into place; it appeared we would be on schedule. The prototype was the only question remaining. But we would soon know, as our flight date was scheduled 15 August 1938.

A Difficult Decision

The modification of old CW-22 to the Turkey Hawk prototype was all done by Stanley Vaughn's sheet metal experts at Curtiss. They had the rear fuselage section deskinned and beefed up in half the time scheduled. You could always depend on Mr. Vaughn and his experimental crew to come through when the chips were down. It was a real pleasure to work with him.

The first flight of the prototype was on time — 15 August — but Lloyd Child, our chief pilot, had to cut the flight short because of an electrical failure in the test instrumentation. The next day, Lloyd stayed up well over an hour. Mr. Gifford and Frank Lakowitz both said, "We have a winner." Child made a typical fast pass over the field, but just a wing over, not a roll — this meant "fair." The roll would have meant good or very good.

In Lloyd's post-flight briefing, he definitely told me to hold up on the production release of the aft fuselage, fin, and rudder parts. He also mentioned that the preliminary stalls felt good and that the rudder was effective through the landing, but added "After the structural integrity test [6.5G], let's take another look. Let's be sure." I agreed, but mentioned that we had a total of 30 days for the total test program and hoped that we could release everything to production in 15 days, or by 1 September 1938.

I reported to Jack Davey and General Andari, and after a thorough analysis of the current facts, they both were satisfied that we could have our first production airplane by 1 March 1939.

The flight-test program on the Turkey Hawk prototype was completed on schedule. Soon after, Lloyd's secretary called from the airport to advise me that Mr. Child wanted to discuss the final results of the flights. I was curious because Lloyd usually wrote the report first and discussed things later. I found, to my surprise, that he wanted me to help in the analysis and report for the benefit of General Andari. When I told him that I couldn't do a good job unless I flew the bird, he responded in his dry way, "I would expect you to do that. As soon as the flight recorder is removed and the standard instruments are installed we'll fly."

Lloyd called me on 21 September 1938. "Let's go flying early tomorrow morning, seven o'clock."

It was a beautiful day. I was at the hangar by 6:00 a.m., ready to go when Lloyd arrived. To my surprise, he asked me to take the front seat. As I checked the bird, I noticed that the spin chute was still installed. This gave me a clue that he expected me to check spins. I felt comfortable with this because Harvey Grey had covered this area with me back in St. Louis. As I climbed out, I thought over a safe sequence of maneuvers: power-off stalls, slow rolls, loops, half-loop roll on top (Immelmann), low-power turn stalls, and power stalls. Everything went smoothly until we tried the power stalls. The bird wanted to snap to the inverted position toward the torque. If this were left unchecked, it could cause an inverted spin. We agreed that this should be tried with forward center of gravity (CG) first. We completed the flight with several power-on, power-off landings. As we were walking back to the office, Lloyd remarked, "You did a good job. Somewhere in your training your instructor taught you all the right ways to do aerobatics. I'm curious who it was."

I replied, "Otto Enderton, a civilian instructor in 1929, and Air Corps Sergeant Pierre DePussen. And, of course, two years flying the weather airplane at Scott Field."

I reported to Jack on the results of my flight and that Lloyd wanted me to write up the report for General Andari. Jack remarked, "What's wrong with Lloyd's people?" Ordinarily, Frank Lakowitz, the engineering representative, would write the report. Frank was in charge of the technical side of engineering flight headquartered at the Buffalo airport.

I told Jack that Lloyd had asked me to do it. He was short-handed, with his assistant chief test pilot, Bob Fausel, doing demonstration work on the P-40 in China, and I simply assumed that I was chosen because I was so close to the program.

Jack understood and reluctantly approved, if I thought it could be done along with my other work. "It might take a few extra hours," I said, "but it will help our department in the long run."

Lloyd Child rechecked the power stall characteristics and finally decided that there was no practical fix other than to put a placard on the panel and a warning in the pilot's manual. This approach was explained to General Andari, and he bought the idea.

All the hard work on my part was over by the end of December. The Turkey Hawk was on schedule. We would have one two-seater first, and about two weeks later the first single-place plane would be completed. The planned schedule was still to have the first airplane by 1 March. In a schedule meeting in the planning department, I saw Ralph Kenyon. He wanted to take me to lunch: "I have something very important to tell you. I know you'll be surprised and interested."

I met Ralph outside the plant at noon; he acted very secretive. I soon found out he had decided to leave Curtiss-Wright at the end of December. He explained that Spartan Aircraft Company of Tulsa, Oklahoma, was expanding and needed a general manager. A friend in Tulsa had called him. He went on to tell me that he had flown to Tulsa to look the plant over and had met Bill Skelly, Captain Max Balfour, and C. W. Gwinn. Skelly Oil Company owned the factory, along with the Spartan school. Captain Balfour managed the Spartan Air College, and Gwinn managed the factory. Ralph also made the comment that his salary was twice what he received from Curtiss. Finally Ralph said that he would give me a similar deal. He visualized my position as the administrative head of the company, except engineering. Earl Weining was the chief engineer. His background was project engineer from Chance-Vought — very well qualified, particularly on Navy projects. The company (Mr. Skelly) wanted to expand the production of the well-known Spartan Executive airplane, a Navy trainer in competition with the Stearman, and an Army advance trainer similar to the now well-known North American AT-6. In summary, Ralph emphasized the necessity to expand the Spartan plant by about 30,000 square feet. We both agreed that we were talking about well over $1,000,000 in up-front money to get the plans underway.

We also agreed that it would be a challenge. On my part, I told Ralph that I would consider his offer, but in my mind, chances were slim because of my family. The grandparents would give me a hard time if I took their grandson away. And I had to admit to myself that things were comfortable in Buffalo — good money, good housing, and living back in God's country with family and friends made for a happy life.

Ralph Kenyon left Curtiss-Wright as planned. Shortly after, production of the P-36 for the French — the French Hawk — was put on a max-effort basis. France was concerned with Hitler's military buildup and his invasion of Poland. They needed a fighter that could challenge the ME-109. Our government sanctioned the French purchase to the extent that they gave the French order priority.

That fall and winter of 1938-1939 was filled with weekend flying — either out of Rochester down to my dad's farm or taking passengers over

Photocopy of a rare drawing of the famous P-36 dive, flown by Curtiss chief pilot Lloyd Child. Note the French insignia. The author received this from a French artist in 1937-1938, on the occasion of the French purchase of P-36s.

Niagara Falls with one or two students from the old Consolidated field. Jean and our son Johnny were with the folks occasionally for a week. The grandfolks had practically taken over our kid.

Howard Burr, one of the loftsmen from Curtiss, was a glider pilot from Elmira. He got me interested in designing an intermediate class sailplane, which gave both of us something to work on in the evenings. He finally boarded with us five days a week in our flat in Kenmore, New York. We went so far as to start work on the machine in our basement, and as my good wife said later, "I wonder how I survived the baby, Howard, and you during the Buffalo days."

In early summer, Jack Davey called me to his office. He did his typical circle talk on how well I had handled the SBC-4, the Turkey Hawk, etc. I knew something serious was going to happen to me. He dropped the shoe. He asked if I would take the parts dispatcher's job on the very hot French Hawk project. I asked him why not the project manager and the stock chasers jobs, too. He replied that Fred Kenney had seniority over me so he couldn't give it to me. To Jack's surprise, I said yes, but I would need my pick of men and 20 percent more money because it would mean more hours. He agreed. I did, however, have Ralph Kenyon's Spartan Aircraft offer in my mind. But the new Curtiss job was somewhat rewarding, as it gave me the experience in manufacturing that I lacked.

I followed the project in about the same way as the last two. Fred Kenney was easy to work with, and the French were interesting, very sure of themselves, and friendly. Their pilots were good people and knew their business. Pete Jensen, Bill Davey, and Jack, I could see, depended on me for the parts flow to make the schedule. Pete Jensen once said to me in private, "I'm glad you stayed. This experience won't hurt your future."

I kept in touch with Ralph Kenyon in Tulsa through all of this. He was slowly getting into trouble with his flying and general administration work and was getting desperate. I finally agreed that I would join him as soon as I could clean up my work on the French Hawk.

My big problem, however, was with the family. The grandfolks were against the whole deal. They wanted to know why: "You have a fine job and good housing. You can fly as you please. Why do you want to tear up your family and move to a strange place?"

I said, "So you can't spoil Johnny every week." My wife was with me. She liked the idea of a new adventure to say nothing of the extra money. We finally cut the cord. I promised to report for work in Tulsa on 15 March 1940.

The hard decision had been made. Curtiss had taught me a lot: men like Pete Jensen, Jack and Bill Davey, Paul Hufgard, Lloyd Child, Bob Fausel,

Jack Kerr, Stanley Vaughn, Don Berlin, Giff Gifford, General Andari, and many others — all my friends. To my knowledge, I had no enemies at Curtiss. It was a part of my life that was a real challenge and yielded much to my career. I remember, at my going-away party, the French pilots toasted me: "Monsieur Cactus Pilote, Bon Voyage."

None of those French pilots survived World War II; they were all killed in early dogfights, defending their homeland.

Spartan Aircraft

Although our folks were very much against the job change from Curtiss to Spartan, they settled down when I promised that Jean and Johnny would come home at least once a year.

Ralph Kenyon was like an old mother hen about the trip, the car, shipping furniture, and the half-built glider. He went so far as to advance expense money for our travels. This proved, to my satisfaction, that he needed me as soon as possible. I planned the trip so that we could stop over one day at Scott Field. The complete route was Honeoye Falls, Buffalo, Cleveland, Columbus, Indianapolis, Scott Field, St. Louis, Jefferson City, Joplin, Tulsa. The total distance: 1,500 miles.

The final parting from our families was a little sad, but the sense of challenge and change softened this feeling. Our interest turned first to the Scott Field visit. Jean had heard the many stories about my stay at Scott. She also had pictures of the operation and was looking forward to our visit as much as I was. We knew there were many changes but were hopeful that the blimp hangar, Operations building, and barracks buildings were still there.

We arrived in Belleville, Illinois, mid-afternoon on Sunday, 10 March 1940. We took a room in the old Square Hotel, then ate in its restaurant. The same people were there that I remembered. This had been a favorite place of Ben Dalley and Dutch Kleinoder, and it was one of mine as well. After we settled in, we drove around the area and stopped at the old Scott Field South Gate. To our surprise, the guard gave me a pass for both sides of the field. This included Operations, the flight line area, and quarters. The old barracks were still the same. The Operations building was in the same area, but much improved.

Across the field, in all its splendor, was the old lighter-than-air hangar. The north doors were open, and we could see a 73 class blimp just inside (73,000 cubic feet of gas, which formed a balloon about 125 feet long). I finally went into Operations. The duty Sergeant and weather briefer showed us around; the layout was the same. I asked about the weather program — there were weather telemeter balloons instead of airplane flights. I also found that Harvey Grey was still on the reserve flying list. The airplanes were 0-47s; the old 0-19s had been retired. The Sergeant talked freely about the lighter-than-air program. The lighter-than-air hangar was going to be removed, as the base was to become the Military Air Transport Command's headquarters, but the Sergeant was not sure what the fate of the 15th Observation Squadron would be. However, the Scott weather center program was expanding.

That evening, I called Harvey Grey and brought him up to date on my new job. The Scott Field visit was a very pleasant experience, but we were left with the sad feeling that lighter-than-air was going to be a thing of the past.

Our interest in the trip turned to Tulsa. The weather was beautiful, in the 60s, and the trip from St. Louis to Joplin was through the Ozarks. The dogwood was in bloom, and it seemed like another world, from cold and snow to sudden spring, with trees like upstate New York. The Lake of the Ozarks was like the Finger Lakes country. Our son, John, asked, "Will Tulsa be like this?" I assured him it would be real nice but deep down I knew it wasn't like this. Tulsa was hot in the summer and cold in the winter, windy in spring, and flat without many trees. We arrived in Joplin at 4:00 in the afternoon and found a hotel room downtown. My "crew" was tired so we turned in early. We departed for Tulsa at 6:00 the next morning and arrived downtown Tulsa at noon. We had lunch and went directly to the Spartan Aircraft Manufacturing plant.

My first impression wasn't good. The main building, including the office front section, was less than half the size of the Curtiss-Wright main plant. However, I did notice that things were neat, with a beautiful front lawn with trees. We drove around the back road toward the airport and around the front of the Spartan school. This was impressive: engine test cells for the engine overhaul unit, hangars and administration on the field side of the main road, and classroom buildings and living quarters on the aircraft plant side — the aircraft plant and the school made up a very impressive layout. After casing the area, we drove into the circle drive in front of the plant. I noted that Ralph Kenyon's Auburn Cord was not parked there.

I went inside and found that the receptionist was a pleasant older lady. I asked for Mr. Kenyon and she asked politely, "Would you be Mr. Rowley?"

They were all expecting us, she said. "Mr. Kenyon will be back in a few minutes. Mr. Gwinn asked that you come right into his office."

Gwinn's secretary was Lillian Mabry. When he came out of his office, he greeted me like a long-lost friend. He wanted to know if Mrs. Rowley and our son were in the car. "Yes," I replied. He told me to "bring them in!"

Jean and little John were taken over by Miss Mabry immediately, and Mr. Gwinn was very much taken with Johnny. We talked about the trip, and to our surprise, Miss Mabry announced that the company had rented a furnished apartment for us, until we could find suitable housing. Mr. Gwinn added, "Be our guest."

Ralph Kenyon arrived. It was like a family reunion. We left Jean and John with Mr. Gwinn and Miss Mabry and made a fast tour of the plant. We met some foremen and the plant manager, and I had a quick flash that there might be bad blood between us when the manager remarked that "all we need is another expert." Ralph didn't respond until we got back to his office, then he apologized for the man's attitude. It seemed the man's family had close connections with Bill Skelly, owner of Skelly Oil and Spartan, and thus was obviously in the company because of this connection. I knew that because of this man's operating methods there was going to be trouble reorganizing the plant systems.

"E. H., this is where you come in. I want you to set up inspection stations for all parts and parts orders, including routing and scheduling." Ralph also commented that tool numbering was already in process. He finished by saying, "We almost have to start from scratch. Mr. Skelly knows the problem because the government civil aircraft inspector — CAA — is threatening real trouble if we don't get a handle on the complete operation." Ralph concluded, "You folks have come a long way. Let's look at the apartment, get settled in, and then we can continue our planning in my office as soon as you are ready. Mr. Skelly will want to meet you in his staff meeting Monday at 9:00. It will be in my office."

The apartment was on the south side in a very good part of Tulsa. It was the answer to our problem: a one-bedroom, totally furnished with linen and towels, all at company expense. We went out for supper, then bought some food for breakfast. Jean and I relaxed.

I spent the weekend writing my proposal for the purchasing, planning, scheduling, and inspection departments. Flight test was to be separated from the school. Spartan Aircraft sales were also to be transferred from the school to Ralph Kenyon, operations manager. Currently Mr. Gwinn was vice president of Skelly and in charge of Spartan Aircraft manufacturing. The school would be a separate operation under Max Balfour.

My outline of operations was partly written to help Ralph Kenyon sell the proposal to Mr. Skelly. I wanted to be chief of the administration side of the operation, of course, including flight test. An administrative chief would be in charge of production planning, purchasing, and parts control and inspection, but not engineering. We had the chief engineer, Earl Weining, for that. I wanted to be chief of flight testing, to approve the flight characteristics of the trainer or any other aircraft.

We had our program review meeting on Saturday, 16 March 1940, at the plant. Ralph agreed with essentially everything that I proposed. This was a good indication to me that he was in a strong position with Mr. Skelly. He outlined the meeting agenda, which included my proposal. He stressed the civilian and Navy requirements for such a reorganization and said, "I'll set you up as the expert. You can be the one to get Mr. Skelly's attention." I realized Ralph was also testing me.

Ralph and I arrived at the plant before 8:00 a.m. on Monday to go over our agenda for the meeting. I agreed to review the civil aeronautics and Navy requirements if we were to go into production on the Executive and the Navy trainer. The Spartan Executive 7W was a five-place corporate-type aircraft, very advanced for its time — single-engine, 200-plus cruise. It was in competition with the Beech Staggerwing but would fly rings around the Beech. The Navy trainer (NP-1) was a primary trainer in competition with the Navy's Boeing Stearman (N2S). The Army version, the PT, was being made by Boeing-Wichita as well. I also agreed to show the suggested departmental layout of the whole organization. The emphasis was to be on the separation of engineering, shop, and inspection.

Mr. Skelly arrived on schedule. Before he sat down, he greeted me warmly and asked about my family and the apartment. I was compelled to thank him and everyone in the meeting for the hospitality and friendly attitude. The meeting was composed of all the top people in the company: Mr. Skelly, president of Skelly Oil; Mr. Gwinn, operating officer of Spartan; Mr. Balfour, director of Spartan school; Earl Weining, chief engineer; Ralph Kenyon, general manager; John North, comptroller; Fred Tolley, shop superintendent; Jay Gentry, chief pilot; and myself.

Ralph introduced me formally and gave my background from military service to Milwaukee Airways and finally to Curtiss-Wright. He emphasized my pilot background and my Navy and other project experience with Curtiss. Mr. Skelly proceeded to outline his plans to release shop orders for 25 Executive 7W aircraft and serious flight testing and development on the Navy trainer. He also revealed that the Navy was expected to release bids for a primary trainer in late 1940 or early 1941. We all realized that the competition from the Boeing Stearman would be fierce. Boeing had built more than 500 airplanes, Ralph Kenyon pointed out. Mr. Skelly remarked, "I know. But I've been told that the Navy wants another source to keep competition alive.

We can get a contract for more than 200 airplanes. We can make it. Our lower overhead is the key." Mr. Skelly, in conclusion, asked Ralph to review the whole program, both cost and manpower, and report in the next meeting.

Our group broke up and Mr. Skelly shook hands with me and added, "I'm glad you're with us. You can be helpful with your background. You know this is a big decision for a small company."

At this point, I felt I was on the team — a feeling quite different from my Curtiss experience. Curtiss-Wright was an old-time company with executives interrelated and tied into each other. An outsider couldn't break through the politics. Interestingly, after World War II Curtiss-Wright was no longer in operation.

Ralph, Fred Tolley, and I stayed over to discuss the program I had worked up over the weekend. Fred was surprised when Ralph outlined the department breakdown. He was upset that inspection, material planning, and purchasing were to be under the new administrative head, and asked, "Who will that be?" Ralph replied, "E.H. has agreed to take the job along with flight test on the Navy project." Fred jumped up and said he would talk to Bill, but Ralph convinced him to "Sit down and listen." He suggested that Fred and I visit about these new plans.

I added, "Fred, I think we can make the plan work. You'll need help when this operation expands by five times. You must know that we'll be forced to double or triple our floor space and our personnel."

Fred asked if Mr. Skelly knew how much this was going to cost, and Ralph replied, "Yes, I've told him 1.5 million. Bill wants to stay in this poker game, Fred. You must know he's lost a bundle by playing with the factory. The school has done well but the factory has soaked up the profit. I can't go along with that."

The Monday-morning staff meeting of 25 March 1940 was one I won't forget. Ralph asked me to review the new department breakdown for them. I diagrammed the proposal on the blackboard, showing engineering, shop, and administration on the same level, all under Mr. Gwinn and Ralph Kenyon. I also explained that this is what the CAA and the Navy would require.

Fred Tolley expressed his unhappiness about having his hands tied, but Bill Skelly listened briefly, then remarked, "Let's not wash our laundry now. I'll see you privately."

Ralph, along with John North, reviewed the financial and personnel requirements, with the exception of the final plant layout and building design. It totaled $1.3 million: new personnel, one purchasing agent, a chief inspector, and one stock room manager. Production planning, from work orders through actual production of the parts, was to be managed by Jim

Lew, who was already on board. Company security and personnel departments were left open, but would be under Ralph Kenyon.

On closing the meeting, Mr. Skelly asked that Dick Hayes introduce me to the Navy trainer prototype. Hayes was the only pilot for the factory — a sales pilot for the Executive and a test pilot on the trainer. Skelly asked chief engineer Earl Weining to monitor this effort, in particular the performance and spins. Skelly closed the meeting with, "Gentlemen, I have the money and you have the brains. Let's make this thing work. If you have any problems, be sure to let Ralph or me know. Good luck!"

Spartan's Navy Debut

I spent the afternoon of 25 March 1940 with Earl Weining. He introduced me to his top people: Fred Stewart, assistant chief engineer and chief of structures; Lloyd Pierce, chief draftsman; and, in particular, two young Oklahoma University aeronautical and math majors — Bob Williams and Gene Reynolds. Fred Stewart mentioned later that they would do the flight test and structural analysis on the Navy trainer. We then spent the remainder of the day in Earl's office as I wanted to check the status of the flight-test program. Earl called in Dave Vaughn, the flight observer who had been on the trainer project from the beginning. The bird had many troubles: overweight, low climb and takeoff performance, and bad spin characteristics. Dave described the Navy ten-turn spin test: "At about the fifth turn, she went flat. Dick pulled the spin chute and it failed. I started over the side. The air flow changed enough that Dick could recover by using power. We had about 1,500 feet left."

Earl then added, "Now that we have a 'go' on the program, I'll make the necessary changes. Hopefully, we can meet the Navy specs without a complete redesign."

I remarked, "Let's not spend a lot of money on a dead horse."

After my meeting with Earl, I stopped in to brief Ralph. I wanted to find out if he knew the true status of the Navy trainer program and found that he was not totally informed — the trainer's problems were a shock to him. I frankly told him that I was against spending money on the old Navy prototype; we should start a redesign without delay. But Ralph wanted me to fly the prototype before making a final decision.

The next day, the 26th, I went to the flight hangar to set up a flight schedule with Dick Hayes. Dick was a typical old-time pilot, one of the best stick and rudder men in the business, though not really an engineering test pilot. Eddie Allen's comment a few months earlier came to mind: Dick was a nice fellow to be around, and a kind of prankster. For instance, he wanted to bet me $100 that I couldn't spin the bird on the first flight. I took the bet, but I didn't tell him about the modifications I had planned before that flight. I went over the maintenance of the bird with the mechanic, Lloyd Brown, and asked for an annual for the aircraft. I would sign it off. I also added that from then on we would do pre- and post-flight inspections according to Navy specs instead of in the loose manner that they had been done.

All special instrumentation would be arranged by engineering and Dave Vaughn, the flight observer, and would be signed off by me. I proposed to Dick a flight-plan form to be signed by Earl Weining and Ralph Kenyon before any flying could be done. In addition, all instrumentation would be listed on each flight plan, along with weight and balance data to be included to control center of gravity changes, which would be critical on all spin tests — in particular, on aft CG limit flights. Dick agreed that a good procedure would be better than the present system.

I spent the remainder of the day first at lunch with Dick Hayes and Jay Gentry, the chief pilot at the school, and then checking and fitting my parachute. We all finished up the afternoon in Jay's office talking about the flight-test programs on the Executive 7W and the Navy trainer. We also talked about the "Zeus" advance trainer and its place in our factory plans. The Zeus was similar to the AT-6 made by North American. I had a good feeling about Jay. He had good judgment and was well versed on the company's problems.

I reported to Ralph about my flight-test plans for the Navy trainer. The modifications included: (1) ballast airplane to a forward CG to 2 percent forward of present figure; (2) add spin strips to top wing, full span; (3) pilot in rear seat only, no observer; (4) mount electric 16mm movie camera on top surface of center section; (5) Dick Hayes to make first flight, a maximum of 15 turns right and left, stick-free recovery; activate camera before entry ("The camera will keep everybody honest," I told Ralph. "If the bird starts to go flat, it will tell us when."); (6) install camera in front seat to show stick position throughout spin test.

Ralph bought the idea and wrote a memo to Earl Weining. He added a flight conference after the results were obtained. Earl was pleased and assigned Dave Vaughn to write the flight plan, instrumentation, and airplane modification.

Dave called a meeting in Earl's office; Dick Hayes, Ralph Kenyon, and I were present. Dave did a good job explaining the spin test, in particular the cameras on the wing and in the front cockpit. To this point Spartan had not advised the Navy on how well the trainer would spin. The Navy actually required ten turns, with stick-free recovery (the "Navy Ten Turns Spin Test"). Dick had never completed the entire test for the Navy, so I was going to. Dave explained that the tachometer, airspeed, altimeter — as well as the stick and rudder pedal positions — had to be recorded. Dick's response to the briefing was negative: "Looks like you don't trust my flying."

I let Earl respond, "Dick, you should remember the near-fatal spin test. We could have used some sound data to determine what really happened." Dick had been flying, with Dave taking notes in the observer's front seat, on that test. A camera sure would have been better. And cameras on the wing and on the instrument panel were our requirements; we had to do them.

I added, "We all know it was a mistake to have carried an observer during such a test, but as it turned out, if Dave hadn't started over the side, we could have lost the airplane, and possibly both you and Dave. We all know afterthoughts are always better. Personally, I would like to fly the airplane before we make any changes. Dick tells me it's very difficult to get the airplane to spin. This is where we have to start. As a trainer, it must spin normally — stick full back to stall and then full rudder to the desired direction. Ailerons should be neutral. I know Dick wouldn't mind a second opinion."

Ralph Kenyon said, "I think it's wise to wait until we have the airplane in test condition. I sense some competition between our two friends [meaning Dick and me]. If I know both, there's some money riding on the test!"

Dick remarked, "Ralph, you know better than that." We, of course, *had* bet on whether or not the trainer could perform the ten-spin test!

We broke up the meeting, all agreeing to fly the bird after the modifications and instrumentation were completed.

The Navy trainer program for the moment was settled. It was time for me to look in on the inspection and planning departments. Ralph and I went to Jim Lew's planning office, and Ralph introduced Jim, department supervisor, and Dale Lawson, his planner. I was impressed with their knowledge. The system they had developed for our expansion was very similar to the Curtiss-Wright system that I was familiar with, particularly on Navy programs — record-keeping and directing of the entire parts production along with proper sequences of manufacture, including tools to be used. With Jim and Dale on board, I felt we would have no trouble.

The inspection department, however, was a real problem. A shop employee working for Fred Tolley was the whole department. The man was qualified, but a "yes" man for Fred. This, of course, would not work in the Navy's plan — for the Navy, inspection had to be absolutely independent from production. This was also one of the requirements of the CAA.

The R-100 glider designed and built by the author and Spartan Aircraft School students. Its first flight was in Tulsa, OK, 27 July 1941.

I had a long talk with Fred Tolley about his department, but we couldn't come to any kind of settlement. Fred insisted that his sole shop employee be the chief inspector under me. I thought that the line of authority was okay, but the man, in my opinion, might still be partial to the shop and not objective and demanding enough. The inspection department had to be impartial, with decisions to be made with engineering know-how.

I felt that it would be necessary to hire an experienced chief inspector and let him hire and train his own people as required, and I reported this in writing to Ralph Kenyon.

I then put the word out that Spartan was looking for a chief inspector and called Harvey Grey at Curtiss-St. Louis and Pete Jensen at Curtiss-Buffalo. I asked if they knew of a good, qualified man, trapped in a lower job, who might be available. The pay would be about the same as a chief line supervisor in the Curtiss plant.

Both Harvey and Pete wanted to know how I was getting along. Pete had heard that Bill Skelly was going to challenge Stearman (Boeing) in a bid for several hundred Navy primary trainers. I told him, "Don't believe everything you hear; but you can't tell about Mr. Skelly."

I wondered how the word got around, and made a mental note to find out. I told Ralph, and he admitted mentioning it to Paul Hovaker from Curtiss in a previous conversation. I think Ralph thought that I was unduly concerned, but I felt uncomfortable about him discussing our business with his old boss.

Our furniture and my glider fuselage finally arrived in Tulsa on 29 March. During our wait, my wife had found a nice house with a garage and shop attached — just what the doctor ordered. The wing design drawings for the glider were completed, and I showed them to Jay Gentry. He suggested that we make them a school project for the Spartan Aircraft mechanics course. I talked to the chief instructor and he was delighted to have a real project for the students to work from drawings to flight. This shortened my glider project by at least one year.

I had talked to some of the Tulsa glider club members and was pleased to learn that two of them had new gliders under construction: one Baby Bowlus, by Gilmore Langworthy, and a new design by Hubert Drake. Their interest helped me to get started on my own, even though my spare time was limited.

Getting settled into the house was good for the family. Jean and I both were pleased with the job, and there seemed to be a lot of potential friends. Even the newspaper interviewed me about my work and my glider project. Tulsa was a friendly place, like a small southern town.

The Navy trainer prototype was to have been completed for the first test flight on 28 March. Dick Hayes made the flight as planned. Stalls, spin entries — both right and left, instrumentation cameras on. The results showed that the cameras worked. The exception was a shadow that was corrected with a 12-volt floodlight on the lower part of the front instrument panel. The first extended spin test was scheduled for 29 March: takeoff 7:00 a.m. Jay Gentry and I flew chase with Mr. Skelly's Spartan Executive. The plan was to start the right spin at 12,000 feet above surface. The flight also required that the spin chute be deployed at the first sign of a dangerous flat spin. The entry into the right spin looked normal and remained so through the sixth turn. Then the nose started to raise and lower in 180-degree increments, and we saw the spin chute. Recovery was immediate; the nose dropped straight down.

Jay and I were at the hangar when Dick Hayes taxied up. We rushed over and congratulated him on a good job. I asked, "Did you turn the cameras on?" He thought he had; but he was a little shook up, which we all could understand as he had had a rough ride.

Thank God, the film was good. It showed the airspeed increasing after the entry. At the point of the spin chute release, the airspeed was at about landing approach speed. Dick commented that the stick pressure was reversing and the rudder was fading — all a sure sign of a flat spin. At the close of our flight conference, Dick said to me, "Old buddy, the next flight is yours."

My question: "Is the bet still on?"

"Hell, no! She goes in good now. The bet is off."

Earl Weining remarked, "Looks like we're going in the right direction."

After a careful study of the film of the last flight, we all decided to further modify the prototype for at least one more spin test. I insisted that we follow the same course: (1) add spin strips full span on the lower wing; (2) move aft CG limits 1 percent forward; (3) increase the fin area about 10 percent; (4) increase the rudder area below the stabilizer. Minimum total area increase for rudder and fin, 12 percent.

Earl agreed to study the proposals. He was concerned about the spin chute that had to be installed on the lower part of the rudder. He off-handedly stated, "We might have some turbulence problems caused by the housing for the spin chute."

I wanted to design a new spin chute installation, attached to the fuselage, not to the rudder. I finally squeezed out a three-week schedule for the next test, which gave me time to interview people for chief inspector, purchasing agent, and head of stock control, as well as time for completion of drawings for the plant expansion.

During a few early morning flights, Dick Hayes checked me out in the original Navy trainer and the Executive. The trainer proved to be exactly what I had expected: low-performance climb and takeoff, but spin entry close to normal. I felt that we had a handle on the spin problem; however, with the performance problem, I thought that the prototype program should be scrapped after the final spin test. We would learn from the old bird how to approach the new design.

The prototype changes were finished on schedule. I made the spin test on 10 April as planned. We went through 12 turns to the right before the flat spin signs started. The nose started to come up about every 180 degrees of the turn, but it wasn't violent. I decided to release the stick and rudder. The spin stopped within two and a half turns. This was close to passable. The left spin was as good or a little better. After the spin, I pulled the bird into a tight loop. As I went over the top, I saw the Executive about 1,000 feet below. It did the same maneuver. I would bet it was Dick Hayes. He wanted me to watch his stuff in the big bird. Dick was smooth as glass. We were showing off, pure and simple.

The film of the spin test was good. It proved that all of the changes were in the right direction. Earl Weining, in our next staff meeting, made his pitch for a complete redesign. Mr. Skelly approved and asked for the flight date. Ralph Kenyon thought we should shoot for 12 months. Mr. Skelly wanted a refined date in our next meeting.

During our building time on the new trainer, I had managed to hire our chief inspector, Ted Wilkins, from the Martin California plant. Ted was a real asset to the company. He had the inspection department in operation in 30 days. I can still see Ted and Fred Tolley standing in the middle aisle of the plant waving their arms and talking. Ted was direct and outspoken; he knew what he wanted done.

During the year, through some political arrangement, Mr. Skelly had a Navy officer, Commander James Knipe, from the Naval Aircraft Factory, inspect our plant. Ted Wilkins was to work with him for our "shakedown" — a Navy facilities check. With the exception of minor problems, we passed. Earl Weining had known the Commander while he was at Chance-Vought, which was a plus. Engineering made a fine show for the inspector.

Our bird was similar to Boeing's Navy N3N. Dave Vaughn and I went over the flight-test procedures for Navy trainers to let Commander Knipe know that we were familiar with the rules. We showed him the old prototype instrumentation cameras, and he wanted to know how the spin test went — in particular, the 15-turn, free-stick recovery. He said with a wide grin, "How many times did you use the spin chute?" Our answer was, "Twice." He told us that the spin test was one of the Navy's most difficult requirements — and the first check a Navy test pilot will make.

In our final meeting with the Commander, some time in April or May, Mr. Skelly asked when the Navy request for bids would go out for new trainers.

Commander Knipe thought that by 1 July of that year and said he would see that we would get the mailing. Mr. Skelly then asked that the forms be mailed to Ralph Kenyon.

We all knew that we had a tough job ahead to complete the prototype so that we could bid on the trainers. The spotlight was on engineering for releasing drawings. But the pressure was off me for the time, so I started work on my glider project. The school was working on the wing, and my shop at home was taking shape.

Little did the Spartan school and factory employees know the changes that would take place in the near future. The 12-month schedule for completing the trainer prototype would be close. I flew the new bird on 29 May 1941.

The NP-1 Project

The initial flights of the redesigned Navy trainer in 1941 showed good results. The spin tests were much improved, as was general performance, climb, service ceiling, and top speed.

Mr. Skelly informed us that the Navy wanted our bid on 200 of the airplanes, plus spare parts, by the first of June. I was made the Navy negotiator at Mr. Skelly's request. He gave me two weeks to be in Washington for the bid opening. My contact in the Navy department was Commander Knipe, our former inspector, and Lieutenant Marriner.

Ralph Kenyon was to prepare the cost bid, and Earl Weining was to prepare the engineering documentation. Privately, Ralph showed disappointment that Mr. Skelly didn't pick him for the contract negotiation. My comment to Ralph: "You're the works manager. Who else could put the costs together? I think Mr. Skelly made the right choice. You also must remember flying is part of a selling job." I suggested that Lloyd Pierce would be a good man to go with me to represent the engineering side and make any bid changes that might be required.

The flight-test program on the trainer — the final data on service ceiling, climb tests, and takeoff and landing distances — had to be completed before the engineering document could be finalized. This meant that with good weather we had to complete the tests in less than 10 days. With bad weather, it might be necessary to extend our Washington date. I put Ralph on notice that the flight-test program could cause a delay if the weather didn't cooperate. This threw him into shock, I guess because he wasn't sure how much time latitude Mr. Skelly had with the Navy. I politely suggested that we clear up this matter before Mr. Skelly fired us both. As it turned out, even with a tornado in our flight-test area southeast of Tulsa, we made our schedule by flying some of the tests at night.

In our final meeting before leaving for Washington, Mr. Skelly asked that I take the American Airlines sleeper that was non-stop to Washington. My reservations were at the Willard Hotel in a two-bedroom suite. I arrived at 4:00 a.m.; the old, noisy DC-3 sleeper was no treat.

Lloyd Pierce was to arrive the next day, so I made contact with Commander Knipe, who set up a preliminary meeting with me to go over the figures and delivery schedule. The Commander was very cordial, but firm. After he looked at our figures, he told me to call Mr. Skelly. "Tell him his bid is high by $3,200. Young man, I will see you at 10:00 a.m. tomorrow."

I called Mr. Skelly and Ralph Kenyon to report on my meeting with the Commander. Mr. Skelly wanted to know how firm he was about the bid being high.

"I can't tell until I try. Would you approve $2,500 firm?"

"Yes, but don't lose the deal."

I took from his comment, however, that he would approve a $3,200 discount if push came to shove. Ralph had no comment.

Lloyd and I talked over our dilemma. We agreed that the company had a learning period coming, and production of the first 100 airplanes would be a good checkpoint. By that time we should know if we were making money or losing it.

We had the new bid ready by the next morning. I made the call to Ralph at the plant, and he bought the proposal and promised to check with Mr. Skelly and call back before 10:00. Our original bid was $13,400 per airplane (less government-furnished property, GFP). The proposed new bid was $11,400 for the first 100 airplanes, with the cost to be reviewed at the 75th unit produced so that the last 100 airplanes could be changed or improved with no further cost change. I made it very clear to Ralph that we wanted to make the bid firm and stressed to Mr. Skelly that we were talking about $240,000 in the company's favor by proposing a 100th-airplane break/evaluation against a possible loss of $640,000 if we tried to produce the entire 200 on a full-steam-ahead/no evaluation break method.

Commander Knipe was trying to compare our bid with Stearman, who had built about 500 airplanes and, "It damn well isn't fair!" Ralph said. He also said he would play the tape of our present conversation back to Mr. Skelly. (It was standard procedure to tape telephone conversations for any future clarification.)

"Okay, Ralph," I said, "but do it by nine o'clock. We have to be back in Commander Knipe's office by ten with Mr. Skelly's answer."

Mr. Skelly called back by 9:00; his answer was firm. "Go for it. If he doesn't buy it, I don't need to lose that kind of money."

Commander Knipe was not as tough as I thought he might be. He remarked that he personally agreed that we had a chance due us, but he couldn't make the final decision. Assistant Secretary Knox would have to review the bid. Commander Knipe excused himself. His secretary was an old pro and remarked, "Relax. He doesn't go this far very often. It shouldn't take long." She offered a cup of Navy coffee that was strong as a bull.

The Commander was back in about one hour with a smile. "Relax, men. The Secretary called Mr. Skelly to okay the deal. I'll have the detail people in my office tomorrow at nine. We can finish up then."

Mr. Skelly called late in the afternoon to tell us the specifics and asked what we knew about the contract papers. Lloyd and I suggested that John North, the Spartan comptroller, fly to Washington with Dick Hayes in the company's Executive airplane. We could finish up the contract work and return sooner and cheaper than on the commercial airline. I had a selfish motive: It would be nice to get some time in the Executive. The idea received immediate approval, and John and Dick arrived the next day. We had our meeting with Commander Knipe at 2:00 instead of 9:00. John signed the contract for Mr. Skelly at about 11:00 the following day.

Dick Hayes and I were at the old Washington airport checking the bird when Lloyd Pierce and John North arrived. We had lunch and were on our way back to Tulsa at 1:00. We stopped at Louisville, Kentucky, for fuel, and were back in Tulsa that evening at 6:30. John had called Mr. Skelly from Louisville, and he and Ralph met us at the airport and we had dinner. North said privately that he had never seen the old man so wound up. He added, "You and Lloyd did a fine job."

The 9 June 1941 company meeting was classic. They all turned to Earl Weining and me first for any engineering changes to get the NP-1 to specifications, and next, for production release orders. Earl had about ten production changes; but no large tooling problems. I explained that our first production release would be 10 airplanes, then 40, then 50, with cost review and renegotiation at number 75. We had enough material to start. We all agreed, and I was to call Commander Knipe to set up GFP schedules (for engines, wheels, tires, brakes, etc.). The Commander was pleased with our progress and asked for the first flight date. Off the top, I gave him 5 September 1941. He advised me that Lieutenant Marriner and Mack MacKine would represent the Navy; they would arrive within ten days.

MacKine had been my old Navy friend back at Curtiss, now assigned to Spartan. It would be a challenge for me to work with him again — he played by the rules, with no variations — but he would be an asset for a young company like Spartan. Our own chief inspector, Ted Wilkins, could handle him. I thought it wise, however, to brief Ted on my past experiences with Mack.

The heat was on for the next three months, with overtime in the plant, training new employees, and buying and stocking new material and new plant building and equipment expansion. Every supervisor worked almost around the clock. We all felt good — no one objected. Mr. Skelly and Mr. Gwinn were pleased and spent a lot of their time at the plant. As I look back now, we went from one airplane a month to one every other day. What a challenge!

The first ten NP-1 airplanes were given the Navy serial numbers 3645 to 3654. This number series was to continue on for the full contract of 200 aircraft. Dick Hayes was test-flying the last three Spartan Executive 7Ws produced, so he let me fly out some of the routine tests for the CAA sign off. I finally realized that he was hoping to get the first flight on NP-1 No. 3645, and I didn't have the heart to turn him down. Dick flew that initial flight on Wednesday, 3 September 1941. Takeoff was at 6:55 a.m., just as the factory people were coming to work. He put on a good show; everyone's morale was at a peak. Later on at the hangar, Mr. Skelly broke out the champagne for his supervisors. Max Balfour and Jay Gentry also were there to enjoy the celebration. Jay remarked, "I just can't miss a celebration with champagne and airplanes!"

That same day I flew NP-1 No. 3648 out of serial number sequence. It was ahead of 3646 and 3647 because of inspection troubles that had forced their delay.

John North had his team in place to study the production costs from No. 3645 and on, including spares. We all were concerned about the 75th airplane renegotiation, and Mr. Skelly wouldn't let us forget. By the time the 75th aircraft was in production, John had a good handle on the cost, and we were breaking even with our present contract at $11,400 per unit. The renegotiation would be a problem, however, as we would not be able to make a cost cut, but we could prove that we were making progress on cost reduction.

But all the effort on production was upstaged by the attack on Pearl Harbor on 7 December 1941. Every male from 18 to 35 was worried. And we were forced to change our purchasing practices as materiel control was by allocation. You had to have a strict reason for everything ordered.

Gene Gubser was a lawyer who became the Spartan purchasing agent. He was a great asset to production planning and was also a great negotiator for the company where legal problems might exist in our government relations.

As we were planning to go back to Washington early in 1942 to negotiate the last 100 airplanes, Mr. Skelly called us all together to announce his loss of control of Spartan and Skelly Oil Company to J. Paul Getty, the New York financier. This news was shattering. Fred Tolley and Ralph Kenyon turned nearly white and left the room. Earl Weining, Ted Wilkins, John North, Gene Gubser, and I stayed on to find that Getty was bringing in his own team to manage the Spartan plant around the first of February 1942.

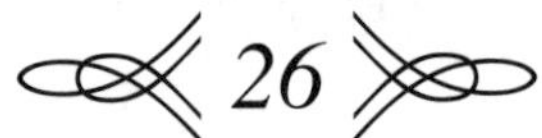

J. Paul Getty

The days before Mr. Getty was to arrive gave us all a chance to evaluate our status. Some of our top people were angry. Some faulted Mr. Skelly, some Mr. Getty. The majority were against Mr. Skelly for losing the company; the remainder were just plain curious about Mr. Getty. I think John North, Spartan's comptroller, in our first staff meeting after Mr. Skelly's demise, put it to us as straight as he could. "Mr. Skelly watched his cash flow but not his stockholders. It was his stockholders who sold out to Mission Oil Company that would shortly be controlled by Getty Oil."

John also mentioned that he had met Mr. Getty once at an oil company executives' meeting in Tulsa about a year before. "He seemed to be mild-mannered and quiet; however, people listened to him." John had been asked to stay on with Spartan in his present position and decided to accept. He told us, "Personally, I feel Mr. Getty will go all out for military business, both for the factory and the school. I know that he has good relations with Admiral Forrestal. Mr. Getty knows his way around Washington."

Earl Weining, our chief engineer, was the first man to come to me about his decision to leave the company. Boeing had offered him the post of chief project engineer of the trainer division in Wichita, and he felt that the possibility of a new airplane would be out of the question for Spartan.

Ralph Kenyon, general manager, was the next. He had a job with a furniture company that was going to build troop carrier gliders for the Army; he was going to manage the plant.

Fred Tolley, shop superintendent, took leave for 60 days, and Archie Barnett, former shop foreman, took over manufacturing. Archie was a top-notch man for the job.

This 1942 photo of J. Paul Getty was given to the author on his last day with Spartan Aircraft Company.

Mr. Gwinn, Spartan's chief operating officer, and Lillian Mabry, his secretary, had taken retirement at the end of 1941. And I kept on working with production planning, purchasing, and materiel control. Flying was fun.

The grandparents on both sides decided to come to Tulsa for Christmas 1941. Jean's father and mother, Ralph and Louise, came by car. My mother and dad came by bus. Their visit was good for the whole family.

The war was going to interfere with Ralph Austin's building business, as supplies were being allocated by the government. For a small organization, it was going to be very difficult. I suggested to him that he come with Spartan to supervise our maintenance department. I agreed to look into the possibilities. Mother and Dad Rowley went home after the first of January 1942. Louise stayed until April, and Ralph stayed with us through November. He enjoyed working at the plant.

Al Reitherman, J. Paul Getty's administrative assistant at Spartan Aircraft Company. Al was killed in 1944, flying an experimental aircraft. The author lost a good friend. (Photo, *Spartan News*)

The information for the NP-1 (Navy trainer) contract renegotiation was ready for Mr. Getty's approval in early February. The Navy had asked that we delay until Mr. Getty assumed his duties with the company. John North called all of the supervisors to a meeting on Monday, 9 February: Fred Stewart, engineering; Archie Barnett, manufacturing; Ted Wilkins, inspection; and me, administration and flight test. We had about 30 minutes with John. He advised us that there was no agenda for the meeting, but thought it would be introductions all around and maybe some questions.

Mr. Getty had brought in Ken Walkey from Douglas as plant manager, and Al Reitherman from his own parts plant in California as his assistant. Our meeting with all of them was low-key; a few remarks indicated that Mr. Getty had been well briefed. In closing, he asked for time to review the operation, adding that he would call us individually as soon as possible.

John remarked after we adjourned, "I think Mr. Getty will eliminate the group meetings in favor of more personal contact. You can also bet that he will work eight hours a day and will spend full time in Tulsa."

In my opinion, Mr. Getty was going to be interesting to work for. His approach was clear to me: The best and only the best would survive. Al Reitherman and Ken Walkey were topnotch men, and that was what he wanted.

On Wednesday, 11 February, after our big meeting, I had scheduled the first flight on Serial No. 3742, the Navy trainer, at 8:00. On takeoff, the wind was from the south over the plant. At about 200 feet altitude, I had a serious engine malfunction. It was necessary to land on the only clear ground just north of the plant on a spot about one block long, *between ditches*. There was no damage to the bird. By the time I got out of the seat, I had company on the way. Mr. Getty, Al Reitherman, my secretary Doris Biers, and Jay Gentry, the Spartan School's chief pilot, were coming across the open field with Mr. Getty's car. Everyone was looking at the airplane for damage. I was shedding my parachute and helmet when Al and Mr. Getty came around the bird with a big smile. Jay walked up and remarked, "How the hell did you get down that short?"

"Well, Jay, it's my hillbilly background in upstate New York. Short fields, you know." This got a laugh, even from Mr. Getty. Lloyd Brown, the mechanic, arrived. "What went wrong, boss?"

"Bad cylinder, I think. A lot of noise."

Mr. Getty took all this commotion in stride. He thanked Doris and said to Al and me, "Let's go back to the office and get together in 30 minutes."

I went back to my office in the plant, and all the people crowded around to wish me well. I was famous for the day.

As I walked into Mr. Getty's office, he stood up and said, "Welcome aboard." Al Reitherman congratulated me on a good piece of flying, and my response seemed to please Mr. Getty: "We didn't damage the bird." He outlined his personnel plans and asked that I continue as his administrative head. I was to receive a substantial raise. In addition, he thought it wise to check out another pilot. He went on to say, "I would like you to head up the contract renegotiation in Washington."

I thanked him for his confidence, then asked Al Reitherman if he would like to go. Mr. Getty, without any hesitation, interrupted, "Max Balfour and John North will accompany you. Be free to fly the company airplane."

Al remarked, "I have plenty to do here." Al was Getty's assistant, and he needed Al to help with all the details there. And Max, as the new Spartan vice president, and John, as comptroller, were both appropriate choices to accompany me.

Almost two hours had passed before my secretary drove me back to the hangar. Lloyd Brown had the trainer engine running when I arrived. The engine trouble was simple. Two rocker arms on two different cylinders were

Max Balfour, the author's good friend and adviser through his Spartan Aircraft Company days. (Photo, *Spartan News*)

loose; the lock nuts were not tight. I remarked to Lloyd, "Find the SOB and fire him! That was a bad show in front of Mr. Getty."

Lloyd remarked, "Yes, boss. My fault. Will fix." Before lunch that day I flew the troubled bird again for what actually would be its first credited flight. I had no problem.

The trip to Washington was great. At 10,000 feet, we had a 55 mph tail wind; our flight time was four and a half hours. The Executive 7W was a beautiful bird. Max remarked to both John and me, "We could build a lot of these if we could get the material and engines. Maybe Mr. Getty will do just that after the war." (We had been forced to cancel all work on the 7W because of World War II.)

Our meeting with the Navy was short. Commander Knipe, who was so firm before, was like an old friend. Our price was the same and he approved our $11,400 bid immediately. Admiral Forrestal asked us to give his regards to J. Paul and added, "We need to talk about some field service problems, both flying and mechanical. Ask him to call me at his convenience."

After we left the Commander's office, John remarked, "Things are different." Max added, "Yes, 100 million different." Mr. Getty received a lot of attention because of his financial position.

Max Balfour, not very well known to me before, took on a completely new image. During our trip he was helpful and outgoing. His comment during our dinner at the Willard the night before we left was, "If I can be of any help, don't hesitate to call either Jay or me." This was a change. I made a mental note. Max had always been rather stiff and precise, but Jay, Max's chief pilot and close friend, obviously had put in a good word for me.

Our trip home was long. We had a head wind from surface to 10,000 feet. I flew the bird all the way.

Mr. Getty was pleased that our bid was accepted. I brought up the disturbing rumor that the NP-1 wasn't doing well with Navy instructors who were asking for inverted spin demonstrations.

John North gave Mr. Getty Admiral Forrestal's regards and request for Mr. Getty to call the Admiral at his convenience. I added that the Admiral had mentioned the field service problem. I could see Al Reitherman was going to get his first big negotiation job. I thought he would have to go to Washington after all to talk with the Navy. I checked with Doris, my secretary. Lloyd had two airplanes ready to fly. As I drove home, I wondered, "How would the NP-1 spin upside down?"

The next morning I checked in with Mr. Getty. We talked about the bad reports from some of the bases that had the NP-1 trainers. He suggested that I call Commander Knipe. "He's an old Navy pilot; he can find where the trouble is."

I placed the call from the outer office with Al Reitherman on the extension. After exchanging proper greetings with the Commander, I asked about the scuttlebutt that we were hearing about the bird not measuring up to training requirements. The Commander came back loud and clear. "Our Atlanta, Georgia, base is the one putting out the propaganda. They're scheduled to get NP-1s only, no Stearmans. I know that they're putting out that the Spartans are bad spinners and have low performance because they want the Stearmans. I suggest that you pick your best demonstration pilot and let me know when you can have him in Atlanta."

Al and I briefed Mr. Getty. His reaction was very firm and very much out of character. "I want this dealt with totally and as soon as possible! I want a meeting with all concerned after I call Admiral Forrestal!"

I called Lloyd Brown to get the prototype ready to fly, complete with spin chute. "Should be ready by daylight tomorrow, boss," was his remark. I also asked Jay Gentry to fly chase for me at 0600. Jay asked, "What's up?"

"I'm going to put on an air show and I want you to watch," I said. "Suggest you fly the Executive. We need to go to about 12,000 feet. We'll need radio contact. I'll operate on our test frequency." Jay went one further. He brought a movie cameraman from the school.

It was a beautiful morning. The climb-out was with the sun. At 12,000 feet I flew the bird inverted for about 30 seconds. The engine was still running. I repeated for 40 seconds. The engine started to miss. I leveled out and talked to Jay. "We can get 40 seconds before the engine complains."

Jay came back, "Looks good. But I don't think it's enough time for a one-turn spin inverted. I know what you're going to try." Jay thought I was going to do an engine-off — let the engine stall — and continue to spin.

"No, Jay, we'll do the simple stuff first. The inverted spins will have to wait for another day."

Jay and his cameraman landed. Doris, my secretary, had called to tell us that a meeting was to be in Mr. Getty's office at 10:00. Jay thought we could have the film in two hours.

"Too late, Jay," I answered. "You saw the show. Tell him. We can do the inverted spins later, after we put an electric pump in the main fuel line to the engine and restrict the wing tank vents."

Max Balfour, Al Reitherman, Jay, and I were present at the special meeting. Mr. Getty confirmed what Commander Knipe had told me the day before. Getty was surprised that we had already made a practice flight to check out an air show plan for the Atlanta Naval Training Station. Jay explained that the show was planned to quell the complaints once and for all. He also said that we would have movies of the final flight routine for his approval. "Mr. Getty, I think you will be pleased with Rowley's show."

I explained that to complete the inverted flight demonstrations it would be necessary to install an electric fuel pump and special tank vents. "We'll design the pump and vent modifications as a kit that we can take with us. I'll check with engineering to get local Navy approval, including flight test. My friend, Mack (MacKine), will be watching."

I could see that Mr. Getty was getting a real kick out of our aggressive approach to the problem. His final remarks were, "I feel confident with the team I have. We, together, can solve any problem: engineering, manufacturing, sales, and flying. I'm proud. By the way, I've had calls from Mr. Jensen of Curtiss and Mr. Schaefer of Boeing-Wichita. They want us to bid on some very substantial work. In closing, gentlemen, I want you to know that I will announce a complete reorganization very soon. All of you present figure into the top management of our new Spartan company."

Fred Stewart and Lloyd Pierce had the inverted flight kit designed,

The author and the "air show" NP-1 airplane No. 3696 at Atlanta Naval Training Base, 2 April 1942. (Photo, Spartan Aircraft Company)

installed, and Navy-approved in time for the first complete demonstration flights the weekend of 28 and 29 March 1942. The inverted spins, as it turned out, were quite easy. The airplane behaved as well or better than in normal spins. This fact made my job easy because very few primary instructors had training in advanced aerobatics. This was my "Ace in the hole." I had to overwhelm our critics from the outset. Jay had filmed all of the practice and, finally, on late Sunday afternoon, called me on the radio. "That's a wrap. The last inverted spins, both right and left, were beautiful."

I saw the final film of the 20 advanced maneuvers at midnight in Mr. Getty's office. The group was complete, including Ken Walkey, the new works manager. Even Max Balfour, the old pro, remarked, "That's damn good."

My final remark to the group was, "The bird will do the job, but it takes a lot of muscle. I think we will have the advantage if we show the film to all the instructors, then invite any that would like to take a ride. This approach will separate the men from the boys. I can be ready to travel Tuesday morning. However, we'll need approval from our local Navy and, maybe, Headquarters in Washington."

Mr. Getty assured us that he would handle this, and added, "Jay will fly the Ex airplane with Rowley and his mechanic." Before Jay could say no, I allowed it was a great idea because Jay had the experience and seniority to handle such company meetings. Mr. Getty remarked, "That's settled. Let's get some rest. I appreciate the effort."

We departed from Tulsa at 8:00 a.m. on Tuesday, 31 March. The weather was 500 feet and 3 miles. I was glad Jay was flying. He climbed to 10,000 through about 3,000 feet of solid clouds. It was a wonderful experience to watch and learn from a professional like Jay Gentry. We were flying at an estimated ground speed of 270 mph. We landed at the Navy base in Atlanta in just over three hours; the weather there was perfect. The Spartan Executive was a great bird.

We had our preliminary meeting with the base Commander that afternoon. Lloyd Brown was given the go-ahead by the chief of maintenance to install the pump kit. The instructors meeting was scheduled for 9:00 Wednesday. I was to fly the show at 2:00 Thursday. I had a feeling that the Navy was going to put on their own show. I had asked Lloyd to check on any rumors in the maintenance department. If any special work was being done, they would know. That evening at dinner, Lloyd reported that everything was quiet. He did say that there was another hangar that was off-limits.

I test-hopped the show bird the next morning, Wednesday, 1 April. The airplane the Navy had assigned was No. 3696. It flew great. I went north of the base to be out of sight to practice. The show Thursday, however, was to be right over the main runway of the base. Jay and I had lunch with the base Commander and the chief instructor. We talked about the NP-1, its good and bad points, and the training problems. In general, there was nothing specific. The Commander did ask about the altitudes that I would be working and safety considerations, in case of engine failure. He was satisfied with spins from 10,000 feet and standard aerobatics as low as 2,000. As we were departing, the chief instructor asked if I would mind if he tagged along with a Stearman N3N. "Our line chief will have a throat mike so that we can talk as you go through your routine. It works good in an open cockpit."

We were at 10,000 feet by 2:00 p.m. Thursday, as planned. I did a right inverted five-turn spin, climbed back, and repeated the same spin to the left with a stick-free recovery that took about two extra turns. I talked to the chief as we went along. The thing that impressed him was the inverted spin performance. Five turns inverted, both right and left. I did an inverted field pass at 2,000 feet to end the show. The Commander and Jay were in the tower listening to the chatter between the chief and me. This communication made for a very interesting instructor's meeting after the air show. The base Commander gave the Spartan team thanks and praise for our efforts. He closed by inviting the instructors to stay for Jay's movie and my question-and-answer session. As it turned out, most of the instructors wanted to know about the inverted flight and spins — in particular, the procedures during entry and recovery. I finally realized that those present had very little advanced aerobatics training. This was a primary training base. These men, however, were the ones doing the complaining.

I had a very interesting telephone conference with Commander Knipe. We heard no more about the bad performance of the NP-1. After we arrived back in Tulsa, David "Deed" Levy, Stearman's chief pilot, called me about our air show. He took it as a sales deal, and his boss wanted to do the same show at their principal bases. I told Deed only what he already knew and asked him if he had worked the Stearman trainer in inverted flight, spins in particular. His answer was noncommittal. I never knew how their show turned out because Deed was nearly killed in Boeing's AT-15. He pulled the wings off in a dive test, but he recovered and retired.

The reorganization that Mr. Getty had promised was good for the company, but not really for me. In February 1942, Max Balfour had been made vice president of Spartan Aircraft, including the school; Ken Walkey became works manager; Al Reitherman, assistant to Mr. Getty and Max Balfour; Ted Wilkins, chief inspector; and Fred Stewart, chief engineer. I had been made chief production engineer. That included purchasing, production planning, stock control, and plant expansion. Max Balfour took over the flying. I did retain field service and Navy contact in the field, however. And the change

Delivery of the last NP-1 manufactured, *The Spirit of Spartan,* a gift to the U.S. Navy from J. Paul Getty. All the production work and labor was furnished by Mr. Getty and the Spartan employees. Left to right: Navy Lieutenant W. F. Marriner, Max Balfour, "Mac" MacKine (Navy inspector), author (in cockpit), and Ken Walkey (Spartan plant manager). (Photo, Spartan Aircraft Company)

was a promotion for me, including salary, but the challenge was gone with no new aircraft planned and thus no test-flying. My main concern would be parts and pieces for Boeing, Beech, Curtiss, and McDonnell.

The last trainer that I flew for Spartan on 10 April 1942 was called the *Spirit of Spartan,* an airplane that the company and the employees gave to the Navy. There was no serial number assigned by the Navy. She was flown to the Memphis, Tennessee, Naval Training Base and was used until salvaged at the end of the war.

My time thereafter included traveling to various companies to obtain subcontract work. We were not planning to build more complete airplanes or to extend the contract with the Navy for NP-1s. The plant expansion required in order to accomplish the work involved a cost of $2,000,000. The various contracts obtained by 1 September 1943 totaled about $4,000,000. This was what Mr. Getty understood best.

Boeing-Wichita

In August 1943 I was in Buffalo, New York, to sell Curtiss-Wright on a Spartan Aircraft contract to build forward cargo doors for the C-46 Commando. I ran into Harvey Grey, and it was like another family reunion. We had dinner that evening, and we discussed Eddie Allen's death on 18 February in the B-29 prototype. Eddie was the "Dean" of all test pilots. He had worked for Curtiss on the C-46 and for Boeing on the B-17, and had done a lot of work at Wright Field as well. Harvey reminded me about our lunch with Eddie a few years before and his talking about test pilots and their technical knowledge.

Harvey's reaction to Eddie's death was a mixture of frustration and anger at the loss of his friend: "Eddie should have been in the tower riding shotgun on one of his top pilots and crew instead of in the cockpit! And his loss puts Curtiss and Boeing in danger of losing both the 3350 engine contract and the B-29 contract." I agreed with Harvey on this — Eddie was too valuable; he should have been making vital flight decisions from the ground instead of trying to bring the crippled aircraft back to the field. On the ground, in all likelihood, he would have told the pilot to take the aircraft, with its engine on fire, out toward the sea and bail out. But, like any good test pilot, he was trying to salvage the machine so that the cause of the fire could be determined, and he didn't make it. The fire burned through the wing, and it came off. The plane with Eddie and his crew of 11 crashed into a meat-packing plant just short of the old Boeing Field at Seattle. All were killed including 29 on the ground.

Harvey was close to Eddie because of the C-46 program. Eddie had been the engineering adviser on the design and had made the first flight two years before in St. Louis. I suggested that Harvey take Eddie's place at Boeing. "You have what it takes, Harvey."

"Oh, hell, I can't do that," he said. "I'm in the middle of the XP-55 program."

The XP-55 Ascender was a Curtiss monoplane, forward-elevator fighter aircraft (thus the "Ascender"). In retrospect, its performance did not turn out too well as it proved to be a slow airplane — high drag. It was never sold to the military.

Harvey was convinced that Boeing had plenty of good pilots to fill the B-29 spot — Slim Lewis, Clayton Scott, Elliot Merrill, and N. D. Showalter on the technical side, but he thought they really had to get on top of things or they'd lose the contract. President Roosevelt had appointed Vice President Harry Truman to investigate the whole B-29 program. And according to Harvey, Curtiss-Wright was going to be in real trouble because of the engine problems.

I also had a few minutes at Curtiss with Pete Jensen. His first words were, "How you doing, Cactus? I hear you're doing well for Getty. He'll give you experience that no one else could. Do a good job on the C-46 doors." As I left his office, I felt a little sad. Pete had a special spot in my career, one that no man could fill. He was a great individual, one of a kind.

When I got back to Spartan in Tulsa a few days later, things were very busy. I spent most of my time on the expansion of the plant floor space. Fred Stewart did the drawings and I let the contracts. I missed the flying, but made up for it by flying my gliders on weekends. I had made the first flight on the R-100 glider on 17 July 1942. It was a joy to operate from the beginning. Gilmore Langworthy and Hubert Drake, members of the Tulsa glider club, were good help to put the bird in the air. Our glider club was a fine group.

My family and I had been enjoying our new home for the past ten months, and I had decided to spend the whole weekend with them over Labor Day, 1943. On Sunday of that weekend, I received a phone call from Harold Zipp, the chief engineer of Boeing-Wichita. On the line with him was Earl Weining, my friend from Spartan, now with Boeing. Earl had left Spartan after Mr. Getty arrived, to take over Boeing's PT and N2S programs at Wichita. Harold opened his conversation by congratulating me on my April 1942 "air show" in Atlanta with the Navy trainer. "I've seen the movie; it was a fine flying job. The Navy has asked us to do the same film with our N2S Stearman. Commander Knipe has given you high marks on your flying ability."

I thanked Harold but added, "I don't think this call is about my great flying ability."

"No, Elton," he admitted, "it's about you coming with Boeing to take over our new engineering flight test department. We've decided you're our first choice. You have both the flying and the administrative ability to get the job done. Frankly, this is a difficult situation. The engines are at fault. You know your way around Curtiss-Wright. This would help. How soon can you come to Wichita to look things over?"

"Harold," I answered, "have Mr. Schaefer call Mr. Getty." Schaefer was Boeing's senior vice president in charge of the Wichita plant. "I have a tough situation here. I'll call you Tuesday evening at home."

Earl Weining had the final remark. "Old buddy, we need you up here!"

The rest of our holiday was at home going over what this change would mean: selling the house, a new school. Jean, my good wife, I think took a dim view. I remarked, "It could be worse. There's a war on. I must do what I can. At least I can be home and stay in the good old USA."

Mr. Getty called me to his office about 10:00 on Tuesday. He explained about Mr. Schaefer calling, and that he would discuss the matter with me and call him back later. "Elton, I need you here, but I understand that this would be a great opportunity for you and it would have a future after the war. What would be the equivalent here?"

My answer was, "General manager and to build airplanes after the war. I should call Mr. Zipp by this evening. I would like, with your permission, to fly up to Wichita and look over the situation."

"Take the Executive if you want."

Al Reitherman heard our conversation. He was in my office by the time I arrived there. "This is an opportunity that only comes once. Don't turn it down," he said. "Boeing is in business to build airplanes for the long pull. I have a friend and relative who is the Army Air Forces representative for Boeing-Wichita: Colonel Ralph Vaughan is a topnotch pilot and engineer. You'll see him a lot if you run the flight test department. It's his business to help settle the B-29 engine problems that are GFP. Remember, the government is really responsible for releasing the engines before they were ready." As Al left my office, his last comment was, "I envy you. This is a great opportunity."

I called Harold Zipp Wednesday morning. "Estimated arrival is ten o'clock. I'll advise your tower on my exact time; I'll have the tower operator call you."

I arrived at the ramp at exactly 10:00 and parked in front of the administration building, just north of the airline parking area. There were several people on the steps — I thought probably waiting for an airliner about ready to pull up. To my surprise, they were waiting for me: Harold Zipp, Boeing's chief engineer; Earl Schaefer, vice president; Jack Clark, assistant chief engineer; Ray Hoffman, head of engineering personnel; N. D. Showalter, B-29 Superfortress project engineer; and Earl Weining, chief engineer of the training project. They welcomed me like a long-lost friend. Earl had been doing his homework on my behalf; this was obvious. They knew a lot about my experience and background.

Mr. Schaefer took us all to lunch at the airport restaurant and reviewed Boeing's total objective as to B-29 production and continued production on the Stearman trainer.

We then toured the Boeing plant by truck. I saw B-29 No. 35 about ready to go to the flight line. Mr. Schaefer and Harold Zipp continued on the tour with me; the others went back to work. We happened to meet Colonel Vaughan, Al Reitherman's connection, on the way back to the administration building, and as I was introduced, it was obvious that there was some tension, particularly between Mr. Schaefer and Colonel Vaughan. The Colonel cut short the niceties and said to me, "Great to meet you. Al Reitherman called me. Hope you decide to make the change. There's a great challenge here. My regards to Al and Blondie. Earl and Harold, I'll see you later."

As we walked down the hall to Mr. Schaefer's office, Harold remarked, "Colonel Vaughan is hard to work with — a real stickler. The Colonel is the key to our flight crew's check-out as well as to military pilot loans when we're short. We also have to get an airplane assigned for our flight-test dog ship." The dog ship was the aircraft we put all the mechanical changes on to be tested — engines, propellers, even a new floor! Harold concluded by saying, "You have the whole story. I want to be sure of that." He had briefed me well and wanted me to make up my mind based on actual circumstances.

Ray Hoffman and N. D. Showalter drove me back to the airport. N. D. was interested in the Spartan Executive airplane, so I took them for a short ride on the deck with a low pass and a roll on the pull-up. I let N. D. fly a few minutes, and he was amazed. "Beautiful! Fast and very stable." I had to get back, so Ray and N. D. both said, "You have the job. Let us know when you can start. We're sure we can meet your salary requirements. Fill out the application and mail it to George Trombold, our personnel manager."

On the flight home, I reviewed my situation. To do everything correctly, I should clear the Spartan building project bid releases, then turn over my job to Gene Gubser, Spartan's purchasing agent, and the Navy field contacts to Al Reitherman. I could leave my family in Tulsa until the house was sold and Boeing personnel services had obtained suitable quarters for us. I radioed Tulsa flight service to call Al Reitherman to meet me in 30 minutes at the assembly hangar, along with my wife. I asked for a low pass north to south with the traffic.

The only aircraft around when I approached Tulsa were a group of five NP-1 trainers taxiing out for takeoff — probably going to Memphis. My pass was low, fast, and noisy. The victory roll was slow. Al and my wife would know that this meant success and to start packing. As I taxied up to the hangar, I saw Mr. Getty, Ken Walkey (the plant manager), Al, and Jean. They all were smiling, so I was sure they had understood and approved of the show. We had an informal meeting in the hangar office about my trip. Mr. Getty was very interested in my comments about the situation — the engines, in particular. He asked, "Did they tell you about the Truman committee investigation of the whole program?"

"Yes, Sir. Harold Zipp leveled with me on all the problems — in particular, the engines. There are no other engines available, so they must redesign and change, which will require a lot of flight testing. This is where I come in. They offered me the job. I did not accept until I've cleared things here with you. Housing up there is a problem. There will be 35,000 people at Boeing alone."

Mr. Getty remarked, "You know you have to do this. It's your duty." His comment was a relief.

We had a formal meeting the next morning — Max Balfour, Jay Gentry, Ken Walkey, and Al Reitherman — and Mr. Getty presided. He repeated what he had said at the hangar. I suggested that Gene Gubser and Al Reitherman take over my work, with the addition that Jay Gentry should do any flying that might be required. "It would help if Mr. Schaefer knew that I had your blessing, Mr. Getty."

"You can rest assured, I'll do just that, Elton."

I cleared up all the work at Spartan and was to start at Boeing on 10 October 1943. Jean would follow in ten days. The Spartan employees and management gave us a touching going-away party. They presented us with wonderful luggage and a beautiful sterling silver, gold-lined, engraved water pitcher, complete with a signed list of "Your Spartan Friends." I had never been associated with a more friendly group; Mr. Getty and Bill Skelly headed the list.

But I was looking forward to being Boeing's chief of engineering flight test. It would be my job to explore the operational and flight envelope of the aircraft and to help work out and flight-test modifications, as required.

The B-29 Challenge

I had asked Mr. Getty for a letter to Mr. Schaefer at the same time I had asked for a current picture for my favorite people album. On my last day, I stopped at Mr. Getty's office to say goodbye. As I entered, he and Al stood and extended their hands, and after some small talk, Mr. Getty turned to his desk and handed me three envelopes, one sealed and one open, and a third, an 8 x 10 manila type. This had to be the picture. Mr. Getty said, "The letter to Mr. Schaefer is sealed with my wax. I have asked Mr. Schaefer not to divulge its contents. The other is 'To whom it may concern.' The large one is my picture. As you'll notice, it is not autographed. I never do." I sensed that Mr. Getty and Al were having more fun than I was. I opened the large envelope. "Mr. Getty! This is your old Oxford identification photo." Getty had graduated from Oxford University, England. This brought out a laugh. It was the first time I had seen any emotion from Mr. Getty. As I remember, I said goodbye with a lump in my throat.

I flew to Wichita the next day, 10 October 1943. I was paged on arrival. A company driver was there to pick me up, and he explained that Boeing had obtained a suite at the McCleren Hotel, just off Broadway, downtown. The Innes Tea Room across the street was recommended for good food. The driver suggested that I leave the hotel at 7:30 each morning to avoid the traffic. I could have breakfast at the plant. "The food is good."

It was almost noon after I had settled into the hotel, so the driver and I went across the street to lunch. We saw Mr. Schaefer with a Chamber of Commerce group. He introduced me around and asked if my quarters were okay, adding, "The employment office is working on a full-size apartment or a house." He suggested to the driver, "Show Mr. Rowley around the east side in my area."

I arrived in Harold Zipp's office at 2:00. I reviewed my application with Ray Hoffman and settled my salary. It was fair and well over Spartan's plan. The life insurance was also very liberal. My pay started 11 October 1943. I mentioned to Mr. Schaefer the sealed letter from Mr. Getty. Harold wanted to know what it was all about, but I couldn't help: "Beats me." I suggested that Mr. Getty and Mr. Schaefer probably knew each other through Admiral Forrestal.

Jack Clark came into Harold's office and asked if I was ready to look over the new flight hangar and my office, lab, and personnel space.

I was ready: "You bet, Jack. Let's get the show on the road."

On the way, Jack reviewed the personnel volunteers for my flight-test department from engineering and shop as well as some outsiders. There were none from Boeing-Seattle. Jack added, "Ernie Allison, the production flight test chief, has the south wing that's finished. The engineering flight test has the east wing facing the flight line. The radio tower is in that wing, also. Harold wanted to leave it unfinished until you arrived so that you can make the layout as you want. You can have office space and drawing board space in engineering until you get your personnel hired and office space completed. I'll have Ray Hoffman set up interviews, starting tomorrow."

The space allotted to me was perfect — lots of windows and from any one you could see the flight line. In the northwest corner of the area was the circular stairway to the radio tower, which was already in operation for Ernie Allison's production flight crews.

N. D. Showalter and his pilot Bob Robbins had arrived from the Seattle plant with No. 1 XB-29. The aircraft was set up for engine testing, which would continue until Wichita was ready to take over the total test program, involving propeller feathering, engine performance, oil cooling, fuel expansion, etc., at altitudes of 30,000 to 35,000 feet.

The B-29 Superfortress was a brand-new, much larger design than the B-17 Flying Fortress, with more than double the horsepower and a curved nose that made it harder to take off and land. The real trouble with the bird was that its Curtiss-Wright engines were brand-new — 2,200 hp giants for the day — and they had had far too little testing for their complexity. But it was wartime, and the engines had been rushed into production and out to Boeing for the B-29 and to Convair for the B-32.

The result, however, was that Boeing was cranking out airplanes that would require serious retrofitting. The company was being blamed for slow production and a lot of accidents caused by the engines catching fire (fires due to oil leaking from the propeller controls and pads that blew back into

Captain Nels Fendrich, the U.S. Army Air Forces Southwest Procurement District test pilot and flight instructor based at Boeing-Wichita.

the hot exhausts). But a slow production record is to be expected whenever there is a quantum leap in technology, as there was at this time. We were actually jumping from open-cockpit fabric biplanes with 400 hp engines of limited fair-weather range, to complicated long-range bombers like the B-29, with the 2,200 hp engines.

I could see that time was of the essence with No. 1 already in the hangar. Bob Robbins introduced me to her, and I sat in the left seat. He asked what I thought.

"Well, Bob, it's like sitting on the front porch and flying the house." He reminded me of that comment many years later.

On the north end, lower floor of the Boeing hangar lean-to was the Air Force flight office. Jack Clark wanted to introduce me around. As it happened, Colonel Vaughan had just come in from a flight, and he was very friendly. He asked me about Al Reitherman and added, "I want you to meet Captain Nels Fendrich. He'll be the man to check out your pilots and will assist you in any flight work you need. Nels, take Mr. Rowley to meet Colonel Harris when he can make it. We also should be talking about a new dog ship for your new test operation. Captain, you organize this with Mr. Rowley."

Bob Williams, chief of instrumentation at Boeing-Wichita. (Photo, Boeing Airplane Company)

Jack Jones, chief project pilot at Boeing-Wichita. (Photo, Boeing Airplane Company)

As Colonel Vaughan left he said, "Let me know when you want to take a ride." I thanked him and told him I'd be ready as soon as I had my feet on the ground. Captain Fendrich smiled and remarked, "You'll have fun; he's one of the best!"

On our way back to the office, Clark remarked, "You and Colonel Vaughan seem to hit it off well. This will help." I told him that Al Reitherman, Mr. Getty's assistant, was Vaughan's cousin and that Al and I were good friends. But, nevertheless, I had the feeling that the Colonel would work me over, but good, on my first flight.

We rounded out that first day looking over applications with Ray Hoffman. I was pleased that we had people to fill all of our primary positions. My idea was to get the top supervision in place and have them hire their crews. As it turned out, Jack Jones (from Curtiss-Buffalo) was selected for chief project pilot, Bob Williams (from the Wichita plant) for equipment and instrumentation, Oliver Mayes (from Lockheed) for technical analysis-flight planning, and Glen Chambers (from the Wichita plant) for chief of aircraft maintenance. I was very fortunate to hire Ann Beaman, an experienced legal secretary from Los Angeles. Ann was a great buffer for me. She could handle a four-star general as well as a tired flight crew after many hours at 35,000 feet and 70 degrees below zero. It wasn't easy for the men to attend a two-hour meeting after a tough flight, but Ann would settle them down and somehow make sense in the conference reports.

Top left: Oliver Mayes, chief of analysis at Boeing-Wichita.

Top right: Glen Chambers, chief of maintenance at Boeing-Wichita.

Left: Ann Beaman, the author's secretary during World War II and the B-29 *Sweet Sixteen* program.

(Photos, Boeing Airplane Company)

I had been with Boeing-Wichita about two weeks when I was paged early one morning to call Colonel Vaughan. It was a beautiful day with broken clouds at about 2,000 feet. I thought this had to be the day for my first B-29 flight. I called back and got the Colonel's secretary who asked me to meet the boss in the Air Force flight office at 10:00. "It's for your parachute fitting."

Captain Joe Drum, one of the military pilots, was there to help me; he acted like a teenage boy about to see his first hanging. Everyone thought I

General Ray Harris, commander of the U.S. Army Air Forces Southwest Procurement District. (Photo, John Harris)

was going to get a real wringing out. The Captain presented me with a beautiful general's grade wool jump suit and cap to match. The chute was new and had a fresh packing date. As I was putting my own private checklist in my knee pocket, Captain Drum allowed, "You won't need that. The Colonel has his own."

My answer was, "I'm sure, but not like this one." I had climb power approach patterns and approach speed. Bob Robbins had given me a thorough briefing, even to the problem in landing because of the curved nose. He also suggested a little power at the beginning of the flare with nose wheel off about 18 inches at touchdown.

Captain Drum agreed this was good information but, "Listen to the boss."

Colonel Vaughan was jolly when he arrived that morning, as was everyone in the flight office: "Good morning, Colonel. A fine morning, Sir."

I remarked, "Okay, guys, let's cut the crap. I know who's the goat today."

The Colonel laughed. "Don't pay attention. They're jealous I didn't check them out. They're Major Rankin's kids." Rankin had been in charge of most all the other pilot checkouts.

I had a good feeling; there was no tension.

To my surprise, the Colonel gave me the left seat. "You might as well start in that position. That's where you're going to spend a lot of time." He took me through the starting procedure and flap position for takeoff. He also showed me his way to taxi, using engines only. I reviewed the takeoff speed from my check sheet against our takeoff weight. He agreed, and I made the takeoff and climbed through a thin overcast at about 2,000 feet, then leveled off at 5,000. The Colonel was hands-off all the way. We spent about two hours in flap and gear-down approach simulations. His main point was, "Be sure to hold your planned speed on all three legs of the approach. I'll show you the reason. We'll do a go-around. Our approach rpm will be twenty-five hundred with Number Six on the boost. Downwind leg, one hundred thirty miles per hour, one thousand above ground. I suggest base leg about fifteen hundred from runway, call for full flap as soon as you turn on final approach. You should be at about six hundred feet at this point, with an airspeed not less than one twenty-five. Okay, E.H., let's take her in."

I guessed that the Colonel thought I was lost on top of the overcast. I had noticed that he had turned on the Automatic Direction Finder (ADF) when we were taxiing out. I said to him, "If you don't object, in the interest of time, I'll make an ADF approach through the overcast."

He remarked, "You're on your own. Work the Wichita radio."

As it turned out, we broke out into the clear about five miles southeast of the field, so I canceled our ADF approach plan. It was fortunate that we were in good position to set up our downwind leg to the north. On final approach, it was obvious, as Colonel Vaughan had mentioned, that it was difficult to hold airspeed. I was a little high over the end. The go-around was good, but I was late on my No. 8 boost setting on the dial and early on the rpm increase order. This was going to take some practice.

On the final landing, things all around were much better. My touchdown was a little long, but on center line and, to my amazement, very smooth. On the way to the office, Colonel Vaughan remarked, "E.H., that was a good job and a good start." I acknowledged that it was pure luck.

The Colonel added, "I got a kick out of your approach through the overcast. I wondered how you were going to handle the problem short of going through a complete approach procedure."

I answered, "We could have been late for lunch if I had gone that route."

The Colonel suggested we go eat with Colonel Harris, who was on the list for brigadier general. "He can be a big help to your operation, particularly in pilot and flight engineer training. The Air Force also has a Link Trainer that your pilots can use to keep up their instrument ratings as well as getting their military green card."

I assured the Colonel that I would put our training requirement in writing as soon as our project pilot, Jack Jones, arrived. "I personally would like to take the advantage of Major Rankin's full transition school." Rankin's school was a 25-hour, all-position program — from engineer to pilot.

As we arrived at the restaurant, Colonel Harris and Major Rankin were on their way in, and Colonel Vaughan introduced me to both. Colonel Harris was very cordial: "I've heard about you from Commander Knipe. In particular, the Atlanta air show." That luncheon meeting started a relationship that lasted until Harris, by then a General, passed away in Wichita many years later.

The propeller feathering program came up in our lunch discussion. "This high-altitude program has top priority from the Wright Field test lab." The Colonel also added that he had assigned B-29 No. 42-26274, a new production aircraft just off the line. He suggested that I check with Harold Zipp who would have the particulars in a day or so. "There probably won't be much instrumentation, so you can get the test started very soon."

On the way back to the factory, Colonel Vaughan remarked, "E.H., you're launched. I also want you to know that this high-altitude program is a dangerous SOB. Captain Fendrich knows about the test; he's ready to go as the project pilot. Major Rankin will go along with a mixed crew. You can get a lot of good experience in the airplane — experience you couldn't buy." I agreed with that comment.

Captain Fendrich and I got our high-altitude physicals. I passed my 30,000-feet chamber ride with a new demand oxygen mask, which was quite different from the old pipestem that we used on the Scott Field weather flights.

Captain Fendrich, as pilot, with me in the copilot seat, made the first flight on the propeller feathering test program on 30 December 1943. Over Wichita we found our 70 degrees below zero needed for the cold-soak feathering test; that saved us a lot of flight time. We were trying to duplicate flying at high altitude for a long period to determine what we had to do to get the propellers to feather. We lost number four engine on the way down and narrowly avoided disaster by slowing and descending to 5,000 to warm up the propeller nose, but we were unable to feather any of the propellers under cold-soak conditions.

The propeller feathering test program took many difficult flying hours using every possible solution, the final being a lightweight special oil pumped into the propeller control cylinder to dissolve the congealed engine oil. All airplanes were retrofitted with special oil tanks and pumps installed in each nacelle. The problem was solved without accident or crew injury, but not without several close calls. Captain Nels Fendrich did a fabulous job of flying throughout the program.

Our test B-29, *Sweet Sixteen*, No. 42-6206, was modified with the new propeller feathering oil tanks, and the system proved successful over the many hours flown on the *Sweet Sixteen* program.

When the new B-29 flight-test facilities were ready on 1 January 1944, Bob Williams, from equipment and instrumentation, was the first man to move in his unit. Jack Jones, our project pilot, arrived about the same time and had his basic crew checked out by early February: Leonard Small, copilot; Glen Chambers, flight engineer; George Harris, right blister observer; and Dave Talbot, left blister observer. The instrumentation and airplane modifications were completed, and the first flight on *Sweet Sixteen* with the new crew was in mid-February 1944. The general flight-test program on *Sweet Sixteen* went on nonstop.

Oliver Mayes, head of the analysis group, and his crew were working around the clock to reduce the reams of data coming off the test instrumentation. This critical data was documented and forwarded to Wright Field's appropriate laboratory. Bob Williams's instrument and equipment group kept the test instrumentation calibrated and in repair and operated the equipment in flight as well.

At about this time, Harry S. Truman came calling, as Chairman of the Senate investigating committee on war production. Boeing managed to stall him by explaining the engine problem, while Curtiss-Wright put on a massive program to remedy the faults.

Curtiss-Wright developed new cylinder heads, new valves, and better cooling to prevent the engine overheating at low altitudes and high boost, and, most important of all, switched from carburetors to a German-style high-pressure fuel-injection system that finally solved the cooling problem. This fuel-injection system provided the exact amount of fuel to each cylinder, therefore eliminating the hot, lean conditions in some of the 18 cylinders.

Testing the B-29 made us pioneers. We didn't work for quick success; that came from steady, solid work, day after day.

On 14 May 1945 Harold Zipp called me to a secret meeting in the Boeing engineering conference room. Jack Clark, Roy Rotelli, Wayne Dalrymple, Harry Soderstrum, and I — engineering management — were present. Harold began, "I was informed that the armament laboratory at Wright Field requires several B-29 high gross-weight takeoffs, carrying exterior-mounted bombs between the inboard engine's nacelles and the fuselage. This is a highly classified project and very urgent. The weights and other data will be given to us by a test engineer who will arrive here tomorrow. Harry and Roy will work with you, E.H., on the bomb-rack installation and flight-test requirements. A special airplane will be assigned for the flight tests. It will be a new plane off our line. I've been assured that the new fuel-injection engines will be installed. This should help you with the high gross-takeoff

Boeing-Wichita engineering management, January 1944. Left to right: seated, Jack Clark, assistant chief engineering; Harold Zipp, chief of engineering; Ray Hoffman, engineering administration. Back row: Wayne Dalrymple, chief project engineer; E. H. Rowley, chief of flight test; Harry Soderstrom, mechanical project engineer; Kermit Thompson, chief of service; "Kip" Kipfer, weight control; Earl Weining, trainer project engineer; unknown; Jim Wickham, chief of aerodynamics; Harold Penix, chief draftsman; Vern Hudson, chief of preliminary design; Roy Rotelli, project engineer; Bob Alred, chief of structures; Cecil Barrow, chief of engineering staff. (Photo, Boeing Airplane Company)

Boeing-Wichita Dept. 600 Maintenance Group for *Sweet Sixteen*. Front row, left to right: Glen Chambers, foreman (flight sevice), Marble Hollman, Leo Bramwell, Milton Hein, Steve Howland, Paul Latimer, Ralph Lupton. Back row: Elmer Blanchat (inspector), James Helems, Theron McPhail, Irvin Bryant, Elwood McKanna, LeRoy Ralston, Lawrence Flander, and Randall Rowland. (Photo, Boeing Airplane Company)

B-29 70060 with Grand Slam bombs mounted under the wings. (Picture taken just after takeoff.) (Photo, Boeing Airplane Company)

Jack Peacock, Boeing-Wichita project pilot. (Photo, Boeing Airplane Company)

cooling problem. I've been told that our top gross weights will be the highest ever carried by a B-29. It's up to us, however, to determine the maximum weight we can handle consistent with our runway length."

The special bomb-test airplane, No. 44-70060, was ready for us on 1 June 1945. Jack Jones and I flew her several times loaded to over 140,000 pounds; that was as far as we could go and be safe. We found that if we started the takeoff roll with 0 degree flaps, and rolled down the runway bleeding the flaps down slowly until takeoff flaps were extended at about the reject point, it would give us the shortest runway distance to lift off. We tested several smaller bombs mounted on each side with no problem. The big test was two 10,000-pound "Grand Slam" blockbuster bombs, one on each side. This would be a close call. We made takeoff using all of the runway with absolutely nothing to spare. That was where I drew the line, for safety reasons.

We finished the bomb test in record time. It was interesting to find, after the war, that the Grand Slam bomb was the standby in the event that the atomic program failed.

Jack Jones and his crew, along with Bob Williams's men, continued to fly *Sweet Sixteen* until the end of the war. Jack left the Boeing Company to head Jack and Heinz Company's flight department in Cleveland, Ohio. Jack Peacock, one of Ernie Allison's production test pilots, took over the project pilot's position after Jack Jones. The basic crew remained the same, except Bob Swiler from Ernie Allison's production flight group at Boeing was

Grand Slam bomb flight crew, left to right: George McVey, flight-test engineer; E. C. Brann, equipment engineer; unknown; E. H. Rowley, copilot; Jack Jones, pilot; and Leonard Small, observer. (Photo, Boeing Airplane Company)

Sweet Sixteen pilots and crews: Jack Peacock, pilot, in copilot's window. Front row from left: R. D. Beckford, test engineer; Bob Swiler, flight engineer; Bob Williams, chief equipment engineer; C. H. Mongold and J. S. Harrison, flight-test engineers. Back row: E. H. Rowley, chief of flight test; Leonard Small, copilot; Dave Talbot, aerial observer; George McVey, flight-test engineer; R. H. McEwen, senior flight-test engineer; E. C. Brann and Warren Radcliff, equipment engineers. (Photo, Boeing Airplane Company)

added as flight engineer.

During the brief time before Jack Jones left, he, Jack Peacock, and I flew *Sweet Sixteen* to get the CAA's unlimited horsepower multi-engine pilot's rating. We then could fly any size aircraft with any power limit and show it on our license.

I, personally, made the last flight in *Sweet Sixteen* on 22 February 1946. As we made the final pass across the field, the ramp was empty. Plant Two had been closed after the war, and I was forced to lay off all of my great people. I had moved to Plant One, the old flight office in the "bay window," a place where all the trainer pilots had been housed back to the Stearman days. Bob Williams stayed on to set up a flight-test instrument lab in Plant One. Ann Beaman, my secretary, went back to the law office in California. I was the only pilot left.

The story of *Sweet Sixteen* was best told in a Boeing company news release:

FOR RELEASE SUNDAY, MARCH 10, 1946

Wichita, Kansas March 9 — This is the story of an airplane — the saga of "Sweet Sixteen." No rows of painted bombs for blasted Jap cities are on the sides of her sleek nose, no rising sun flags for Jap fighters shot down. Yet to this one plane goes just as big a share of victory as that carried by the great squadrons of her sister bombers thundering over the vast expanse of the Pacific.

During the war, this story couldn't be told — she was a plane that had "Secret," "Confidential" and "Restricted" stamped all over her. She, her flight crew and her ground crew were a mystery to most of the workers at Wichita Division plants of Boeing Airplane Company. But now, that story can be told.

"Sweet Sixteen" was No. 42-6206, the 16th airplane of the total of 1,644 giant B-29 Superfortresses to leave the big double doors at Boeing-Wichita Plant II. The Army bought her and sent her to Smoky Hill Army Air Field at Salina, Kansas and then recalled her to perform experiments never before undertaken by any B-29. She had a good background for her job, too, as she had been a test plane at Smoky Hill.

She arrived back home November 8, 1943, and immediately was assigned to the Boeing-Wichita engineering flight test department. From the fuselage came all gun turrets, all surplus equipment, all armor plate that would be used in a regular warplane. Into the fuselage went over three miles of extra tubing, twenty-five miles of special wiring; into the rear bomb bay went a floor supporting great racks of pressure-indicating manometer tubes. The entire seating arrangement was changed for test personnel.

Along her sides and top, and even on racks supported by the

floor, went panels containing scores of gauges and indicators, a Brown recorder to keep a constant check on temperatures, a wire recorder to preserve the oral records dictated by the flight crews, a photo recorder and batteries of movie cameras.

And into the front bomb bay for possible emergency use of the crews went the first pneumatic snap-opening bomb bay doors to be installed on any B-29. This occurred long before the electrically-operated doors were replaced by snap-openers on production Superfortresses.

In short, the B-29 "Sweet Sixteen" became a flying laboratory performing at home the tests and experiments that enabled other Superforts overseas to show Japan the real meaning of air power.

During the many months from late 1943 to early 1946, "Sweet Sixteen," under the supervision of E.H. Rowley, chief of the engineering flight test unit, and flown by such experienced Boeing pilots as Bob Lamson, Jack Jones and Jack Peacock, rolled up more than 535 hours in the air — mostly at high altitude, even to 40,000 feet.

While carrying out this long experimental program, "Sweet Sixteen" was powered by 26 different 18-cylinder, 2,200 horsepower Wright Cyclone engines. These great power plants consumed 440,000 gallons of 100 octane gasoline and 12,000 gallons of oil. And while they were pulling the B-29 up to 40,000 foot altitudes, the instruments and cameras inside were taking down the facts the Army Air Force needed in its offensive against Japan, as many as 15,000 separate readings on one flight alone — this flight lasting seven hours. And flying the airplane, watching over the delicate instruments, and noting mentally their own observations were the trained men of the Boeing-Wichita engineering flight test department, in their own way just as much a part of the Pacific as . . . the boys at their positions in combat Superforts.

"Sweet Sixteen" was a trouble shooter all right, a necessary part of the Twentieth Air Force which for the first time in history of war was sending against an enemy nation great fleets of swift, pressurized, long-range, heavy bombardment planes — designed, engineered and built by Boeing for the sole assignment of crushing Japan.

There was the time, for instance, when a hurry call for help came from the 20th Bomber Command in the China-Burma-India theater. The extreme temperatures there were causing precious

The total flight-test group at the end of the *Sweet Sixteen* program: 47 men and women who did the impossible in 28 months. The author is seventh from the left, middle row. (Photo, Boeing Airplane Company)

gasoline to boil and gush out the vents of B-29's, especially after attaining 20,000 foot altitudes.

The CBI temperature conditions were simulated at Boeing-Wichita where the crew used a steam-heating apparatus to pre-heat 3,500 gallons of gasoline, then pumped the fuel into "Sweet Sixteen's" tanks and took off for high altitudes to note the reaction. The tests were repeated until the engineers found a method of preventing loss of fuel — a method immediately adapted successfully to other Superforts in the C-B-I.

Then the control cables started giving trouble. At high altitude, the temperature ranges down to as much as 85 degrees below Fahrenheit and the metal skin of the B-29 shrinks, this shrinkage in the wings amounting to as much as one and one half inches. But control cables did not shrink that fast. Thus, with slack controls, the bombardier was deprived of a steady platform from which to release his explosives. After tests in which hydraulic tighteners failed to work correctly, the engineers solved the problem by inserting into the cables a metal corresponding more closely with that used in the fuselage.

Of "Sweet Sixteen's" 16 major experiments, four of the most important were carried out at high altitude. They were the power plant performance test, requiring 135 hours of flying and eight months for completion; the propeller feathering test, 130 hours; the oil cooler and inter-cooler performance, 80 hours; and the ADI, or water injection test, 71 hours. And then there were tests for oil tank pressurization . . . , C-1 autopilot, fuel tank electronic fuel gauge, propeller anti-icing, wing and nacelle pressure, battery vent, cockpit glass, general cooling and others.

Tedium, physical and mental strain — even exhaustion — were the conditions under which "Sweet Sixteen's" crew worked so that the boys at B-29 bases could carry on with safer equipment, more efficient and effective equipment. They were checked by Wright Field's best medical officers, and underwent thorough examinations by Boeing doctors every three to six months. These tests were complete — they had to be because every man was a specialist in his own field, had to perform his job.

There was danger in what this flight test crew did, plenty of it. Three times while at extreme altitudes the gunner's sighting blisters blew out and the airplane decompressed immediately. Split-second usage of emergency oxygen masks saved the crew members.

> Another time, during a propeller feathering test, the governor failed to take hold and the prop on the left outboard engine ran wild. Rising to a screaming crescendo, the tips of the 16½ foot blades were soon exceeding the speed of sound. Those at the Wichita airport heard the scream from high above that day and knew what it meant, but they couldn't see the airplane because "Sweet Sixteen" was on an altitude test. Only the quick action of pilot Jack Jones brought the bomber and crew back safely. He put the plane into a stall and feathered the wild propeller.
>
> And now, after 28 months, the B-29 phase of Boeing-Wichita's engineering flight test program is ended. "Sweet Sixteen" has performed the task to which she was assigned — performed it well. There is not a blemish on her safety record — not an accident, not an injured person. In a few days the Army will call for "Sweet Sixteen" and fly her away. When that happens the B-29 Superfortress era at Boeing-Wichita will come to an end.
>
> But in "Sweet Sixteen's" place are other planes — new models, models still on paper, and older models undergoing modernization. With the return of peace, Boeing-Wichita's engineering research and experimental flight test program enters another phase, assumes a more important place than ever before. Where "Sweet Sixteen" left off, new and even better things have begun.

Before closing this chapter, I'd be remiss not to mention Dr. Jim Beaver, our flight surgeon, mother confessor, and dietitian, as well as Jim Crawford, our altitude chamber operator. Dr. Beaver taught us how to fly at high altitude using our decompression chamber at Wichita. Every crew went in there, and he took them to 50,000 feet, then made them take off their oxygen masks. He showed them what they had to do at that altitude in order to survive: hold their breath and grab the emergency bottle and switch it into the system. This was absolutely essential training, and it gave people confidence.

Jim Beaver and Jim Crawford were both very patient with the flight crews. We learned how to avoid the bends and became aware of the "RGE" factor — the Rate of Gas Expansion — in our diet! I was proud of our resultant perfect safety record, something rare in those hard days.

Post-War Boeing-Wichita

After World War II the old "bay window" office was, in a way, a relief from the nonstop work, flying, and meetings that went on during the B-29 program. The plant went from over 25,000 employees during the war to about 400. All that was left was top management and supervisors. Every one of us had big ideas; all wanted to build airplanes. Boeing-Seattle had the B-50 contract in the works and there was some preliminary design in Wichita on a 20-passenger feeder line twin-engine aircraft, but my assessment of the future wasn't good. It would be at least a year before there would be any flight-test work.

I decided to take the family on a vacation in 1946, back to the Rochester, New York, area. I had heard that Bell was going in for helicopters in a big way and so contacted Bob Stanley, the chief engineer. He put me in touch with Joe Mashman, Bell's chief helicopter test pilot. He took a lot of time with me, gave me several hours of dual flight instruction, and offered me a job. The family was all for the deal, but for some reason I held off until we went back to Wichita.

The first Monday after I arrived back, Harold Zipp called all his supervisors to a special meeting to announce that the feeder line design had been approved for engine selection. Flight test was to build the engine test stand and run the test. Bob Williams was our man to do this ground testing. The next deal was a flight research program to develop a new liaison airplane for the Army, the XL-15. The initial wing and control development was to be done using the L-6 Interstate aircraft, a single-engine monoplane, already in service. Earl Weining was made the project engineer, which pleased me. Earl was great to work with on a flight program. It was like going back to

An aerial view of the Boeing-Wichita Plant Two, looking northwest. The small Plant One is to the north. The city of Wichita is in the distance. The picture was taken after Plant Two closed 69 days after the end of World War II. (Photo, Boeing Airplane Company)

the Spartan days. Harold finished the meeting by also announcing that the Nationalist Chinese wanted to purchase the drawings, tooling, and 26 aircraft of the original PT-13D single-engine biplane trainer design. The 26th aircraft was to be equipped with a new Lycoming flat six-cylinder engine, rated 260 hp. We were to first purchase the aircraft from the military surplus storage area in El Reno, Oklahoma. I was to make the selection and arrange for the pilots to fly them to Wichita. Then they were to be completely overhauled at Plant One, test flown, boxed, and shipped to China. Wayne Dalrymple was the project manager. Suddenly the flight-test department was back in business. My thought of changing jobs faded, much to our parents' disappointment.

I knew my way around the El Reno storage area because I had purchased my Consolidated Vultee BT-13A large single-engine aircraft from the same place — the same Master Sergeant — a few months before. I had paid $1,000 for the plane, fueled and checked for the flight to Wichita. I hoped that I could do as well on the Boeing PT-13A trainers for China, but the search for them turned out to be very different. They had been culled down by crop-duster people until I barely found the 26 that we needed. They eventually would be completely worked over, but my worry was flying them back to Wichita. I was forced to take some with poor fabric and rust on steel tubing, among other deficiencies. There were also some engine problems. The Sergeant agreed, however, to make them all flyable. The price: just under $3,000 per airplane.

I made arrangements to have Jim Greer, a local Wichita pilot, fly five men per trip — a little crowded but not too bad — in his Stinson Gullwing. This aircraft proved to be an excellent choice, and Jim was glad to get the business. He told me later that I had saved his airplane from repossession.

I had assembled the crew from the Boeing-Wichita plant: Jim Wickham, aerodynamicist, MIT graduate; Mel Wheeler, chief pilot of the old trainer group; George Hanna, ex-line inspector on B-29s; and Bob Elliott, structural engineer. I was the fifth man.

Harold Zipp had flown me to El Reno to pick up the aircraft that was to be equipped with the 260 hp engine. He was as disappointed as I was about the condition, but agreed it was the best we could do. If we had been three months earlier, the quality would have been much better. Sergeant Becker had my special airplane ready and gassed for the trip back. The weather was clear, 50 degrees with a 15 mph tail wind. We were on the ground in Wichita 1 hour 45 minutes later. We flew low all the way, and I noted that there were good emergency landing fields the entire trip, though there was a lot of standing water. The Salt Fork River, west of Lamont, Oklahoma, was two miles wide, and flooding was very bad in that area.

The L-6 research airplane, forerunner of the L-15. Note the high horizontal stabilizer and trailing airfoil flaperons. This airplane had top surface roll spoilers. (Photo, Boeing Airplane Company)

The first four flights ferrying the PT-13As to Wichita were uneventful. The weather was cold, but clear. All went well until the last flight of five. We were forced down at Enid, Oklahoma. A light, wet snow left the airplanes covered, and they had to be cleaned off as the weather was turning cold. We all got started and were off for Wichita at 3:00 p.m. on 14 March 1947 — not a good day for me. The flight of five was at 1,000 feet over the surface when a frontal passage caused a rapid drop in temperature. Jim Wickham came alongside, gave me the thumb "down to five," meaning drop down to 500 feet. As I leveled off, the Salt Fork River was below.

The north side of the associated flood area was about 500 yards ahead. At that instant, my engine quit cold. No warning. I stalled the bird in, about 50 feet from the edge of the water in a very soggy wheat field, and nosed up and stuck the dead prop in the mud. I crawled out of the cockpit and fell into the muck, chute and all. By that time the others were making passes to see if I was okay. I waved the best way I could to say, "Get home before it gets dark." They understood and disappeared north.

The area was flat so I could see a roof far off, probably a mile. With my chute, mukluk boots, and other gear, it seemed to take forever to walk the distance. To my disappointment, it was an old abandoned barn, but it was near a muddy country road. I knew if I walked east there should be a north and south main road. There was one, three miles later! It was paved, and just to the north a few hundred yards was a house with a pickup truck in the yard. This was a beautiful sight.

As I approached, there was a light and a front porch just like back home. I threw down the chute and fell flat on the deck. About ten seconds later, a small boy opened the door. He screamed, "Daddy, Daddy, there's a hurt man on the porch!" The father came running. "My God, you're a flier. What happened? Where's the ship?"

"Back yonder, about three miles. I was alone." This, I could see, helped his frame of mind.

"Take off your boots and come in. I'll bet you all could use a shot of good old Redeye. Also, sir, the missis will have supper directly. We would like you to stay. You can call from here."

It was a pleasant experience, and Jim Calley became a good friend, for many years. Every time I had the chance, I would buzz the house and sometimes drop things for the kids.

Harold Zipp, Polly, his wife, and Jean picked me up later that evening in Lamont, Oklahoma, at about 10:00. By that time, with a few drinks at Lamont's best watering hole, I was ready to relax, but not until I had told the whole story one more time on the way home. I called the other guys on the flight that night before I turned in.

The factory finally retrieved the downed Chinese trainer from the Lamont area about 45 days after I had landed. The crew that dismantled the airplane advised me that the carburetor heat control was disconnected from the heater box. Therefore, I probably had severe carburetor icing, causing the engine failure. Another lesson learned the hard way.

The Lycoming-powered Chinese trainer was ready to fly in record time, but the first flight was brief. I was shocked with the performance, and to add to my disappointment, the engine cooling was also very poor. These problems caused many extra flying hours. With the XL-15 Scout high-performance aircraft research program and the L6 Interstate program, I could see that there wasn't enough time in the day to meet the required schedule. The answer was to find one more experienced test pilot. Mel Wheeler, chief pilot for the trainer program, was my first choice, but he didn't take the job because he was about to be made shop foreman on the Chinese program. However, he agreed to help fly the production flight acceptance tests on his own time, just to keep his hand in.

George Hanna was my only immediate possibility. He was working as a flight instructor for a small operation on the northeast side of Wichita. I approached him on a weekend when he was busy with one student after another. That was his complaint; he had nothing to do during the week. George was still unhappy about his earlier layoff from Boeing after the B-29 program was finished. I assured him that I had work for at least one year and that there might be a possibility for a permanent job in flight test if he could hack it. He finally agreed to my proposition as soon as he could clear his student list. George came back to Boeing in 1947 and retired many years later from the flight test department. He had made the right decision.

At about this time, rumors were circulating around the military and our political representative in Washington that we were in a so-called Cold War with the Communists — Russia and China. All the aircraft manufacturers as a result were called upon to go back to the drawing board to develop new and advanced aircraft with more speed, altitude, and range. The jet engine was the answer. It had very good thrust-to-weight ratio, though it was very high on fuel consumption. In-flight refueling, a new concept to the industry, turned out to be the only solution to that problem.

In May 1947, the Wichita chapter of the Institute of Aeronautical Sciences elected me president. As a result, I was invited to New York for the National Engineering Seminar. On the agenda, Air Force General Bill Irvine was to give a paper on long-range flight. The goal was for a bomber flight around the world, nonstop. This was an interesting subject because the General was doing his long-range flight tests, using hose-method refueling, on a new Boeing B-50, a modified B-29, with Pratt & Whitney 4360 28-

The XL-15 "Scout" — a publicity picture taken in front of Boeing Plant One. (Photo, Boeing Airplane Company)

cylinder engines rated at 3,500 hp. His paper stressed fuel mixture control, minimum rpm, and high cylinder pressure (BMEP — brake mean effective pressure). These, with the newly developed in-flight refueling, would make a nonstop round-the-world flight possible. The B-50, I could see, was marked for the job, and the General confirmed this to me during a lunch with him the day after his speech.

Boeing-Wichita's flight-test work on the L6 Interstate was completed and the Army gave us the go-ahead on the XL-15: 2 test airplanes and 12 YL-15s for initial production, fully equipped with radio, floats, skis, "Brodie" gear (for landing on a wire in jungle conditions), and aircraft tow gear. I

The XL-15 (picture taken for an Army L-15 “Scout” brochure). (Photo, Boeing Airplane Company)

The XL-15 on Boeing Lake. It flew well on floats. (Photo, Boeing Airplane Company)

The author in the front cockpit of a PT-13, about to check out Colonel Hun Chan Tang, the Chinese representative on the trainer project. (Photo, Boeing Airplane Company)

agreed to fly Phase One, which included all performance testing as well as demonstrations on floats, skis, "Brodie," and aircraft tow. This was going to be very interesting and challenging for sure.

Around this same time, Colonel Tang and his crew arrived from China for the PT-13As. His people — pilot, engineer, and inspector — were sharp. They spoke good English and all had been trained in the United States. Our Chinese program was going well except for the special trainer with the Lycoming engine. With several propeller changes and a lot of testing time, I still couldn't make the performance requirements because of low thrust. The high rpm needed to get the horsepower advertised had resulted in low propeller efficiency. Colonel Tang released us, however, because they had selected the engine, and the airplane was shipped to China with the other rebuilt trainers.

The first XL-15 Scout, No. 6520, was ready for flight on Sunday, 13 July 1947. I was the pilot that afternoon. There was a big crowd, including the engineering department, with some wives, my own wife and son included. On the climb-out, the engine head temperature went up into the red. I pulled the throttle at about 300 feet and leveled off, but the engine would not maintain cruise power without serious detonation. I landed without much damage to the engine. The exhaust gas jet pump cooling was not working in flight. This raised our flight-test hours by 25 percent as engine cooling was very important to an airplane that had a speed range of from 35 to 110 mph.

The Phase One flight test program on the XL-15 prototype was well along by the end of May 1948. The cooling problem had been solved and the performance tests, including takeoff and landing distance, had been passed. Max effort landing over a 50-foot obstacle was 300 feet, zero wind. Takeoff was 500 feet over the same obstacle with zero wind. Another very important problem, however, was rate of roll at slow speed, which was solved by lift spoilers operating in addition to the "flaperons." This combination gave us good lateral control down to and through the stall.

The final problem on the XL-15 that was very difficult to solve was propeller selection. We tried them all. Jim Wickham, chief aerodynamicist, and Verne Hudson, chief of preliminary design — two very wise gentlemen — developed a test that was perfect for checking propeller thrust from static to flight speed. It was accomplished by towing an automobile behind the aircraft. In the tow line was a dyno-scale giving the thrust of test propeller in pounds. The automobile brakes controlled the speed. Readings were taken each 5 mph increment from static to 50 mph. The airplane was in flight at 45 mph. We also tested the airplane drag by towing the airplane through takeoff and well into flight. We did these tests in the early morning to get zero wind conditions.

The McCauley forged aluminum fixed-pitch propeller design, made in Ohio, was selected over all the variable-pitch designs. Our test procedure was a surprise to the propeller designers, as few people in the business could believe that a fixed-pitch propeller could have more thrust than a variable pitch. But the McCauley propeller made our performance objective possible.

Jim Wickham, Verne Hudson, and Earl Weining were great people to work with. Their abilities and inventiveness made the XL-15 test program a joy for me; I had jumped from a high-pressure program like the B-29 to one that was down to earth, but still challenging. Flying was back to pure pleasure.

One of the great experiences I had in between all this was flying the XL-15 at Fort Riley, Kansas, demonstrating the little bird to U.S. Army General Dwight D. Eisenhower. J. Earl Schaefer and the General had been classmates at West Point, and I enjoyed a good briefing from Mr. Schaefer before the flight.

"General Ike likes small airplanes. He flies PT-13s and L-3 Cubs every chance he gets. Give him your best show. I've told him about the airplane and you and about some of the antics you pull around here."

I had flown a PT-13 to Fort Riley the day before to check the layout of the field. There was a parking ramp north and south on the west side of the field with airplanes parked on both sides, leaving a clear strip 150 feet wide and 1,000 feet long, and no obstructions on either end. I got permission to do my landing and takeoff show from that location. All I needed was about a 15-knot wind from either north or south.

Harold Zipp and I arrived early in the morning of 28 July 1947. The wind was south at 18 — just what the doctor ordered. The General arrived about an hour later. I made a few passes to burn off fuel. It was a perfect spot. The air show was brief and close in. All maneuvers were at 500 feet above the ramp, and the course was about 1,500 feet long. The first takeoff was less than 100 feet. I climbed out to the 500-foot level in less than 1,000 feet, then turned to land in front of the General with a ground roll of less than 200 feet, off again, then turned downwind to do the rare attitude change without change of altitude. A 25-degree nose up to 25 degrees nose down. *Still level flight.* The L-15 was the only airplane that could do such a maneuver. The final performance was the square turn, six G vertical, 150-foot radius. Another maximum effort, landing in front of General Eisenhower. He was pleased. He walked over to the airplane as I got out. His first words were, "Good show. That kind of flying takes practice. This airplane is no Cub."

I answered, "No, Sir." I then overheard the General as he was talking to Harold Zipp. "This airplane would take a lot of pilot training to get the full potential." It was obvious that Harold did not expect that kind of reaction.

Florene McVey, the author's secretary during the turbulent postwar period.

Nor did I. I think it was a case of over-cooking the demonstration, but Mr. Schaefer told me later that I did the right thing.

I was feeling relaxed at Boeing about this time. All of our flight programs were on schedule, and the Chinese contract was about completed. George Hanna and Mel Wheeler were doing most of the production acceptance flying, and the Chinese representatives were easy to work with. Colonel Tang was at my home several times; he was a favorite with my son, John, and he missed his family in China. We kept in touch with him for several years after his Boeing visit.

The XL-15 Phase Two flight program was at hand. Captain W. E. Fleck was assigned to Wichita to do the test on No. 2 XL-15. During his flight check, I realized what General Eisenhower was talking about that day at Fort Riley. Captain Fleck initially had a difficult time making the performance test required. After hours of briefing and instruction in flight, however, we made it through. Captain Fleck was not favorable in his report back to the Army, and I had to finally admit that the little bird was difficult to fly, particularly when it was necessary to take the airplane to its limits to meet the required Army performance.

The flight-test technical data was growing at such a rate that it was necessary to hire a full-time secretary to control and manage the documents released for the XL-15 and the special China trainer, model A75L5, final test data reports. The engine test-bed data that Bob Williams was completing also presented a problem.

By 1948 I had needed an office manager. George McVey, one of my B-29 flight test engineers working in Plant One inspection, suggested that I interview his sister-in-law, Florene, who was working for Beech Aircraft at the time as a secretary. Flo McVey turned out to be the perfect answer to my problem, particularly when the flight-test department was about to face its greatest challenge.

OPERATION DRIP

On 12 March 1948, Harold Zipp, Boeing-Wichita's chief engineer, had called a meeting of all his department managers. The subject was "In-Flight Refueling." He advised us that Cliff Leisy, a Boeing-Seattle mechanical designer to be transferred to Wichita, was in England studying a hose method that was being sold by In Flight Refueling, Ltd., to the Royal Air Force. Cliff was a highly experienced project engineer who was subsequently in charge of all in-flight refueling development — pre-, experimental, and final development. The English hose method was thought to be operational for smaller, slower aircraft, like the Handley Page Harrow. Boeing, however, had been asked to come up with its own refueling method as soon as possible for larger and faster aircraft such as the B-29 — like in less than ten days! There were many questions. Mine was, "Where can I get the flight crews?"

Harold said, "Check with Captain Lyle Freed." Freed was the Air Force plant representative (AFPR) for Boeing-Wichita. In 1947 the U.S. Army Air Forces had become the U.S. Air Force. It would be necessary to use mixed crews, composed of Boeing and Air Force personnel, but Harold wanted me to direct our side of the flight operation.

Cliff Leisy was to be back with some British engineers in a few days, and we would be moving our engineering back to Plant Two from the smaller Plant One. Harold told me to activate my old office in the experimental flight hangar and meet with Captain Freed to get two operational B-29 birds for the project. Flying this program was to be our key to success. We were to use any overtime necessary — this was top priority.

"This project is the biggest challenge we've had since the B-29," Harold added.

OPERATION DRIP, test flown by Major Gust Askounist and the author, the first successful flight, 28 March 1948. This picture shows the ideal tanker-receiver position. (Photo, Boeing Airplane Company)

The first test flight of OPERATION DRIP took place on 27 March 1948, with Air Force Major Gust Askounist as pilot and me as copilot. The test was a failure. The next day, 28 March, OPERATION DRIP became a reality.

Following our successful flight and throughout the next few weeks into April 1948, we continued to work on the in-flight program. We knew that the proper in-flight position of the two aircraft was critical.

Captain Lyle Freed and I called General K. B. Wolfe at Wright Field about the B-29s that were being used there for refueling flight position studies. He suggested that we both come to Dayton to observe the flying that was being done by Majors Askounist, Guy Townsend, and Bob Cardanas, Air Force test pilots. "The two B-29s will be available by Saturday, 17 April. We can talk about crews when you get here."

Captain Freed, Captain Stultz, and I flew to Wright Field from Wichita on Wednesday, 14 April. At Wright we went aboard the B-29 that was used for maneuvering around the companion airplane to find the best flyable position for refueling.

My original report on the test follows:

BOEING AIRPLANE COMPANY
WICHITA DIVISION
INTER OFFICE MEMORANDUM

FROM 60- E. H. Rowley 21 April 1948
TO 10- H. W. Zipp cc: Leisy
SUBJECT: Flight Refuel Position Tests — B-29 Aircraft

1. At Patterson Field, Friday 16 April, two flights were made to determine the optimum position and air speed for the subject project. No. 1 flight was made at 5,000 feet with a takeoff gross of 123,000 and 125,000 lbs. tanker and bomber, respectively. The second test was made at the same takeoff gross weight, altitude 20,000 feet. Several different positions were tried at 180 and 200 miles per hour IAS [Indicated Air Speed], with result that 180 IAS proved too slow to hold good formation at both altitudes. At 200 miles per hour IAS, formation could be held satisfactorily as the ship's controllability appeared normal.

2. In summarizing these tests, it appears that there is only one formation position that would work satisfactorily with the present British system. The position of the tanker would be about 40 to 50 feet above the centerline of the bomber, wing overlap 30 feet,

and the tanker pilot's cockpit even with the leading edge of the bomber fin; that is, approximately 20 feet overlap nose to tail. This position probably is short on fuel head [height of tanker aircraft in reference to the receiver aircraft], however; good formation is difficult to hold if a higher position is established as downward visibility is very limited. The 40 or 50 feet height was picked primarily from a visibility standpoint and represents the maximum upward position.

3. The outstanding position as to fuel head was the rearward and upward position; that is, the tanker 75 to 100 feet high and directly to the rear 30 to 40 feet of the trailing edge of the bomber rudder. This would be a reasonable position to fly if adequate visibility could be provided. The possibility of a pilot's position in the present bombardier's seat location providing mechanically similar flight controls is suggested. Rudder pedals, wheel posts, standard throttles, and arm rests would be required. Air speed, bank and turn, dual tachometers, and manifold pressure gauges would probably be required for instrumentation. After trials on this mechanical installation, the possible installation of electronic controls for rudder elevator and aileron could be developed with uncoordinated individual servo-operated controls; the pilot could accomplish better results than could be expected from a formation stick as the rudder and aileron coordination could be maintained by the pilot allowing skidding and slipping which is necessary in formation flying.

4. The next position with outstanding merits is the rearward and downward position; that is, the tanker airplane in the forward position and the receiver airplane or bomber rearward. It was found, in this case, the receiver airplane could be brought up to a position to about 20 to 30 feet below the centerline of the forward or tanker aircraft and approximately 10 to 20 feet to the rear. This position is very easy to fly as the forward airplane's movements can be detected early so that corrections can be made immediately. Of course, this position is completely different and would not be adaptable to the present British system. It is, however, recommended that some investigation be made, in order to determine if the purely American system could be developed around this position. There would be no position change for the pilot necessary as adequate visibility is now available in the present standard aircraft. It is my understanding that a civilian in the Power Plant Laboratory at Wright Field has patents on such a system; that is,

The top view of the 28 March 1948 OPERATION DRIP flight. Note that the tanker-receiver yawed about 5 degrees, and required full rudder trim. (Photo, Boeing Airplane Company)

nose to tail with a telescopic tube to transfer the fuel and a tow line that is attached to the nose of the rear airplane and the tail of the front airplane. In flight, the cable is connected between the two airplanes and the telescopic tube is then contacted to the receiver aircraft and, therefore, the receiver aircraft is essentially in tow.

5. The preceding three positions discussed are the only ones feasible; however, several other positions were tried with limited success. Pilot's observation reports from Majors Askounist, Townsend and Cardanas will be forthcoming.

E. H. Rowley

[Note: The memorandum was initialed by Jack Clark, Harold Zipp, and J. Earl Schaefer and noted for file #20413.]

The story of OPERATION DRIP is best told in a news release published in late 1948, written by Rex Harlow, our Boeing *Plane Talk* newsletter editor.

EARLY DAY BOEING AERIAL
REFUELING STORY DRAFT

Over the radio came a voice from the B-29:

"The hose is fifty feet from the receiver — forty feet — thirty — twenty — ten —."

A pause, then:

"Hose is inside the receiver!"

Another pause, this one lasting for six long minutes. Then, at 6:27 p.m., the shouted words:

"Water! Water!"

At that moment, in the waning hours of Easter Sunday of 1948, "Operation Drip" went down in United States Air Force records as marking a new concept in aerial warfare.

It shrank the vast expanses of oceans and diminished the world's land areas. It reached out and drew the most remote targets within bomb sight scope of Boeing B-29 and B-50 Superfortresses.

On the success of "Operation Drip" hinged a vital phase of U.S. Air Force strategic planning. It formed part of the foundation of the new national security program. And tied directly to its accomplishment was the AF's directive calling for complete reactivation of the huge World War II B-29 production facility at Boeing Airplane Company in Wichita.

High above the plains of Kansas on that Easter Sunday, 1948, droned two Superfortresses built in the days this nation was at war with Japan. They moved in close to formation, one slightly above and to one side of the other. It was a crucial flight for the Air Force and Boeing crews of the two B-29's. Weary but determined, these men had three abortive attempts behind them — two as recent as the morning of that same Easter Sunday.

In the ears of the crew members drummed the words of an Air Force general who only four days earlier had stated:

"We don't care if you have to do it with a teaspoon, but we want an immediate demonstration that fuel can be transferred from one B-29 to another — while the planes are in the air!"

And these same fliers knew that the AF's top planners were counting heavily on the success of their refueling experiments. Just the day before, on a special plane trip to Boeing-Wichita, Arthur Barrows, undersecretary for air, Lt. General Howard Craig, Major General Gran Gardner and other officers had inspected the equipment and had gone over the test procedures.

Behind the "Operation Drip" demonstration taking place that day lay a period of two weeks of preparations and planning in which every step came under the heading of "urgent." On March 15, Lysle Wood, chief engineer for the Boeing Company, who was assistant chief engineer at that time, had called Harold W. Zipp, then Boeing-Wichita chief engineer, to report that a B-29 modification program was "in the mill" for Wichita. Boeing's office at Dayton, Ohio, nearby headquarters of Air Materiel Command [AMC] at Wright-Patterson Air Force Base, had called in the same information.

Then came word that this modification program involved, among other things, installation of air-to-air refueling equipment in the Boeing Superfortresses. Also, that C. J. Leisy, who was to become Seattle's project engineer on the program, was even then in England studying refueling methods developed by In Flight Refueling, Ltd., a British concern.

By March 20, Seattle engineers were arriving in Wichita to assist with the project, and two days later Wichita's Engineering Division began moving from the company-owned Plant I to take over part of the huge wartime Boeing Plant II, which had stood silent and virtually empty since the time of the war.

Up to this time details of the program remained vague. A few preliminary drawings of the British system had arrived and were being studied. It soon became obvious, however, that the British equipment, in its then-present stage, was not adequate for the Air Force plan involving the big Superfortresses.

On March 24, a Wednesday, Brig. General Horace Shepard, AF procurement chief from AMC, arrived with Boeing President William M. Allen to discuss the new project with J. E. Schaefer, Boeing vice president and general manager at Wichita. On the preceding day two B-29's at Wright-Patterson AFB had practiced, for the first time, a "cross-over" method of making flight contact for possible air-to-air refueling.

During the Wichita conferences March 24, General Shepard talked by telephone with Major General K.B. Wolfe, then director of procurement and industrial planning at AMC, and they agreed that the two B-29's would be flown to Wichita on the following day — Thursday, March 25.

Replacing the telephone receiver, Shepard turned to Allen, Schaefer and the others attending the conference and issued his call for an immediate demonstration of B-29 aerial refueling. The general was well aware that he was handing the toughest kind of a job to men who until the last few days had scarcely given thought to inflight refueling of the big Superforts — let alone prepare for it.

But this was urgent! In a way it took on the aspects of a second "Battle of Kansas," recalling that wintry March of exactly four years before when nearly 600 key employees were pulled from their Wichita jobs and rushed to four Kansas B-29 bases to help get the "Wolfe Project" Superforts ready for the initial bombing of Japan.

The activity at Wichita didn't cease with the end of the shift that Wednesday. Roy Rotelli, project engineer, was preparing to leave for home when he received this summons from Chief Engineer Zipp: "We have a real job! Let's get started!"

It wasn't altogether a surprise to Rotelli. Together with A. G. (Bill) Arundale, engineering laboratory chief, he had been conferring with Seattle engineers on the general aspects of inflight refueling. As a starter, Zipp suggested using a hose and reel from a portable fire extinguisher for the first tests.

Armed with wrenches and screw drivers, an engineering crew

headed by Rotelli and Arundale dismantled a fire cart located in the Plant II Flight Hangar. The reel contained 150 feet of hose. Meanwhile, it had been determined that at least 250 feet would be required for the first flight experiment. Scouting among wartime equipment remaining in the big bomber plant, three more hose reels were found in the Camouflage building and two others were spotted on a fire truck which was up for sale as surplus.

Working far into the night, engineers prepared preliminary sketches of the installations to be made in the two B-29's from AMC. By morning the drawings, together with a supply of makeshift materials, were in the hands of the experimental shop and the engineering laboratory. It was decided to substitute water for gasoline in the initial tests, and a standard B-29 bomb bay tank was selected to carry the "fuel" in the tanker airplane.

All day Thursday and most of the night, then all day Friday and until well after midnight, engineers and mechanics labored to install the systems. On Saturday morning, the third day after General Shepard had called for a demonstration, all was in readiness.

Manned by Air Force test pilot Major Gust Askounist as tanker pilot, Captain Lyle Freed as receiver pilot and Boeing crews, the two Superforts took off and made contact. From the lower, or receiver, airplane a cable with sock attached began streaming out to the rear. From the upper, or tanker, plane another cable dropped through a hole cut in the bottom of the fuselage and started moving downward under the drag of a pear-shaped weight.

The lower bomber held to a steady course while the tanker maneuvered until the cables crossed. The tanker's cable slid rearward along the receiver's cable until it reached an automatic grapnel which locked them together.

By improvised hand reel, the tanker crew drew the cables upward until the grapnel was inside the fuselage. Disconnecting the cables, the crewmen attached the end of the receiver's cable to the hose nozzle and the procedure was reversed — the receiver reeling in its cable and unwinding the hose from the reel in the tanker.

But this first test, although nearly successful, was a "dry run." Just as the nozzle reached the receiver, a weak spot in the hose gave way under the strain and the hose flew off into space.

The two Superforts were flown back to the plant. Engineering and experimental crews went to work rigging up a second hose assembly. On Easter Sunday morning, March 28, both planes were ready for the second test.

They took off shortly before noon. Just as on the first attempt, contact was made and the cables connected. Everything worked fine until, with the hose fully extended, the hauling cable in the receiver suddenly parted. Again the hose flew off into space.

The strain was telling on the Air Force and Boeing crews. They were working against time. They were showing the fatigue that results from maneuvering the big B-29's into the precise positions needed for these crucial tests. They were pioneering an experiment that was completely new for airplanes of that size and speed. But once more they rushed the B-29's back to the plants where work had already begun on a third hose assembly. In record time it was installed and the Superforts were off again for the third trial — the second that day.

Moving into position high over the plains of Kansas, the tanker began releasing its contact cable. With 75 feet outside the plane, the cable jumped the reel and became entangled with the hose!

Despite the hopeless outlook, the planes did not turn back. They cruised along for more than an hour while crew members and Boeing engineers worked at separating the hose and cable, then pulled in the cable by hand — an inch at a time — and rewound the reel. Meanwhile, the sun was sinking farther and farther into the west and approaching darkness threatened another postponement.

Back at the plant, Schaefer, other management staff members, engineers and technicians were grouped around a radio in the electrical laboratory, listening to the conversation between the two planes.

With barely enough daylight remaining to complete the test, the process of making contact was once more resumed. Over the radio the tense listeners heard the reports of the hose being pulled into the receiver, waited through six long minutes of silence, then were thrilled by the words, "Water! Water!" shouted by Major W. P. Maiersperger of AMC, the Air Force commander of the flight.

Instead of transferring the "fuel" — the water — directly into the receiver's tank, it was simply piped through the airplane and allowed to flow free overboard. It was this "fuel" streaming back past the Superfort that caused Major Maiersperger to shout the two words that spelled success.

The refueling contact was maintained for ten minutes and the results of the test — recorded here for the first time as "Operation Drip" — are now history. Leisy returned from England the next day — March 29 — bringing with him sample British refueling equipment and accompanied by British engineers who were to act in an advisory capacity.

Then followed months of testing — improving — perfecting. More B-29's were taken out of "mothballs" and flown to Wichita, followed later by new B-50's produced at Seattle. Some were stripped of armament and equipped as fuel-carrying tankers. Others were modified as receivers, or bombers. The big plane filled the huge assembly bays in the reactivated Plant II at Wichita and overflowed onto the aprons and ramps. Millions of dollars in machinery, other equipment, and materials poured into the factory; hundreds of employees were added to the Boeing-Wichita payroll. Both inside and out, Plant II began to take on an appearance and tempo reminiscent of its World War II B-29 production days.

"Operation Drip" had proved that fuel could be transferred from one B-29 to another while the planes were in flight.

The equipment of In Flight Refueling, Ltd., of England had already proved successful in operation, but not to the extent desired by the U.S. Air Force in refueling planes of the size, speed and altitude of the big Boeing Superforts. Boeing and Air Force specialists immediately went to work testing component units of British equipment for familiarization purposes and to determine if they could be adapted to the capacities and needs of the Superforts.

Since the British equipment was designed for smaller-type aircraft, it was apparent that the operating loads would have to be doubled in a B-29 installation. Of paramount importance was the fact that it must be made to work in extreme cold, and at great heights — and that a stepped-up rate of fuel transfer would have to be attained.

The majority of equipment in a B-29 is basically electrically operated, while the British equipment is hydraulically operated. In order to provide large and varied hydraulic fluid volumes, it was necessary to design and install an independent hydraulic system. Another change involved the formating and initial contact procedures. Following tests with the British harpoon gun and projectile method of contact, this was discarded in favor of the cross-over system used in "Operation Drip."

Laboratory tests in which refueling hose lines and other equipment were subjected to extreme temperatures resulted in a rush call to American manufacturers for a new-type hose — one that would withstand the most frigid conditions.

In an isolated area on Boeing property, but well away from the two factories, company engineers set up "Operation Gusher" to determine (1) the power required to triple the flow rate in transfer of fuel, and (2) to ascertain and correct the points of pressure loss in the various components of the system.

A long series of aerial tests followed installation of equipment in "teams" of B-29 tankers and receivers. Boeing flight crews under the direction of E. H. Rowley, chief of engineering and flight test, and Major Gust Askounist, Air Force test pilot, flew mission after mission in checking and perfecting procedures and equipment prior to the first actual transfer of fuel.

Admittedly, it was rough going in the first phases of the program. Crews — both Boeing and Air Force — were not accustomed to flying the heavy and speedy bombers in the tight formations required for refueling contacts. Lengths of cable and sections of hose broke away during the continuing series of experiments. At one time the Associated Press carried this story under an Emporia, Kansas dateline:

> "A Lyon County wheat field has produced something more than the owner expected. Carl Maddon, the owner, described the yield as a copper contraption weighing about 40 pounds with about 475 feet of wire cable attached. He expressed the belief the parts had fallen from a plane."

Farmer Maddon didn't know the significance of his discovery. Neither did other Kansans who found pieces of hose and cable in their fields and pastures. But on December 7, 1948, the anniversary of Pearl Harbor, the Air Force in Washington removed some of the secrecy by announcing that a Boeing B-50 Superfortress, refueled three times in flight, had completed a 40-hour, 9,400-mile non-stop flight from Texas to Hawaii and returned with a "substantial bomb load." At the same time it was revealed that refueling equipment which made the flight a success was installed at Boeing. *(From News Bureau, Boeing Airplane Company, Wichita, Kansas.)*

The B-50, the best and last great piston-engine bomber. (Photo, Boeing Airplane Company)

31

013 on April Fool's Day

In the midst of our in-flight refueling studies, just after our successful OPERATION DRIP flight, I had received a call on 30 March 1948 from my good friend Colonel "Pappy" Hatfield, the Air Force representative for the Boeing-Seattle plant. His first words were, "Old buddy, I need a favor. I need you to check out Captain Freed and Colonel Jim Bradley in the B-50. I'm flying to Wichita. I'll check you out as first pilot and your man Chambers as flight engineer. I'll arrive about 1600 today. Hope you don't mind the serial number — 013. Lyle Freed will off-load for storage some stuff that's on the bomb-bay platform. I'll check in with you when I arrive. You and Gust keep up the good work on the refueling project."

We were to be getting B-50s in Wichita for the in-flight refueling program and, of course, would need to have our pilots checked out in the aircraft, so Pappy figured he'd solve the flight checks and his own needs at one time.

My response was, "Okay, Pappy, what's really coming down?"

Pappy said he'd tell me when he got there. "Promise!" Pappy always had some kind of deal in the works, so any time he said something like, "Hey, buddy," I knew something was cooking.

Colonel Hatfield arrived that day on schedule. The B-50, No. 013, was a beautiful bird. Captain Freed and I met the Colonel as he came down the ladder. His first words were, "Okay, men, I'm going to War College at Maxwell Field and then to Korea. I have a B-29 wing promised. I want to fly your B-25 to Maxwell to check out my quarters."

Pappy had carefully figured that he would check me out in the B-50 and then I, in turn, would check out the others while he was flying on down to Maxwell for his own purposes in our B-25. And actually, this wasn't such a bad idea; it saved a lot of time and expense by getting everyone checked out at one place.

Captain Freed told Colonel Hatfield that his plan would be all right, if we could fly the B-50 to Wright Field in a couple of days. He needed to pick up some British hose reels. Colonel Hatfield thought that this would be fine. He planned to check me out the next day.

It finally dawned on me that Colonel Hatfield had scheduled me for check-out on 1 April — April Fool's Day! Glen Chambers, my engineer, came to my office about 9:00 and informed me that he had checked the bird and she was ready to go with fuel for about three hours safe. I told my new secretary, Flo McVey, that Colonel Hatfield would be calling about my check-out flight. "I'll be on the line with the B-50."

We had barely finished our walk-around when Captain Freed and Colonel Hatfield arrived in a staff car. The Colonel was raring to go. He gave me the left seat in the aircraft; he took the right. Glen Chambers was assigned the flight engineer's position. The Colonel's flight engineer rode the jump-seat.

The start-up was the same as with a B-29: engines 1, 3, 2, 4. They were smooth as glass. But the taxi out was much different; the tiller steering eliminated steering with engines and brakes. At the end of the runway, the Colonel briefed me on takeoff procedure.

"Use the tiller steering to takeoff speed, trim neutral, hands off the tiller steering, pull the wheel back, and you're flying."

I heard the Colonel say over the intercom, "Stand by to be amazed!"

He was right. The big bird jumped off the runway. I pulled the power back to best climb, 45 inches manifold pressure, 3,000 feet per minute. About that time I saw number four tachometer go to zero. The Colonel had feathered the number four engine. That was his April Fool's Day prank. We leveled off at about 10,000, and he pulled power back to cruise on three engines.

The Colonel asked, "How do you like the rudder boost? If you hadn't looked at the tach and throttles, you could have missed the whole thing."

I agreed. "It's a great improvement over a B-29. This bird flies better on three engines than the B-29 does on four."

We shot three landings and all were good. Pappy said, "Let's go home."

The next day, 2 April, Captain Freed and I flew B-50 013 to Wright-Patterson to pick up the British hose reels while Colonel Hatfield took our B-25 to Alabama. I made an instrument approach with a 500-foot ceiling. She was a beautiful-handling bird. Captain Freed and George Hanna piloted 013 back to Wichita for our flight. My total flight time in B-50 No. 013 on 1 and 2 April 1948 had been 6 hours 30 minutes.

On 12 April, Colonel Jim Bradley arrived at Wichita from Dayton to get his B-50 check ride in accord with Colonel Hatfield's plans. Colonel

Dale Allison, Boeing-Wichita liasion engineer (flight test). (Photo, Boeing Airplane Company)

Bradley was a West Point classmate of Pappy's, so before Pappy went to the War College at Maxwell he wanted Bradley checked out. Bradley and Pappy were very close friends and did everything they could to help each other. To me this made some sense. However, this whole check-ride caper had not been cleared with General K.B. Wolfe, head of new aircraft procurement, probably because he would go by the rules and complicate the matter. I was told about this a few days later by Earl Schaefer. The whole plan had been "engineered" by Colonel Hatfield, "my old buddy," but I stood up for him to my boss, Harold Zipp, and to Earl Schaeffer, who questioned me about the arrangement after General Wolfe had called them. But Pappy had managed to have Captain Freed, Colonel Bradley, and me checked out in record time and at minimum expense. I would have had to have flown commercial to Seattle to be checked out by Boeing pilots using new airplanes from production. That would have taken at least a week. Mr. Schaefer acknowledged that I had acted properly, and said that he would work out the problem with General Wolfe.

Colonel Hatfield came back to Wichita from Maxwell Field a few days later and stopped by my office to tell me that he would be in Korea within three months. He wanted me to fly the military's C-46 Commando, a twin-engine aircraft, an overgrown DC-3, to Maxwell with some cargo, which turned out to be his personal items that had been aboard the B-50 013 when he flew her to Wichita. I checked the load in the C-46. The first thing I saw was a baby crib; Hatfield was moving his family goods to Maxwell. I agreed to fly the C-46 if Major Warren, Colonel Hatfield's aide, would go along and file the military flight plans to make it all legal.

But Warren wasn't a pilot, so Dale Allison, my right-hand man and pilot, and I made the flight on our own time over a weekend. My secretary, Flo McVey, remarked later, after Colonel Hatfield had left in 013 for Seattle, "I'm glad the Colonel has departed. Maybe we can get on with our normal business of operating the flight-test department. Hatfield, on top of Askounist, would be too much." Both Colonel Hatfield and Gust Askounist were similar personalities.

During those B-29 days I had the privilege of knowing three very special young career officers — Colonel Pappy Hatfield, Colonel Joe McCoy, and Colonel Jim Bradley. The sad part was that there would be no more "Hatfield and McCoy" capers, like their mock fights in various and famous watering spots around the country. Colonel Hatfield was killed on a bombing mission in Korea and Colonel McCoy was killed in a B-47 accident off the end of the main runway at McDill Field, Florida. Colonel Bradley was seriously injured in an automobile accident near Wright Field, Ohio, causing early retirement from the Air Force. In my opinion, the Air Force lost three great men, and I lost three great pilot friends.

The Flying Boom

Eventually there were four aerial refueling systems. The first was the British hose and grapnel in which a hose was lowered from a Handley Page Harrow flying above an Empire flying-boat. The flying-boat let out a grapnel from the tail and when it grasped the hose, the refueling line was pulled into the tail of the "boat." In another version of this system, the receiver fired a harpoon sideways to come around ahead of the nose to snag it. But at 250 mph, the drag swung it backwards too quickly. This came before the Boeing system.

The second refueling method was the Boeing development in which the recipient let out a hose; the tanker then flew a diagonal course, crossing over the hose, grappling it, and hauling it up into the tanker. This was a much simpler method as the pilot of the tanker could see the recipient flying lower and to his left.

The third system to be developed was the flying boom in which a rigid "tube" was flown by the controller lying in a window area in the after part of the belly of the tanker, guiding the boom into a receptacle in the top of the bomber coming up for fuel.

The fourth system was the probe-and-drogue, which saw the refueling hose with a perforated cone, after its end reeled out from the tanker with the recipient to be refueled plugging its probe into the cone. The main problem with this system was that the tankers were not nearly fast enough for jet aircraft. Probe-and-drogue constant-drag drogue was developed by a small company south of San Francisco, after the boom system, primarily for USAF fighters and especially for U.S. Navy fighters.

At Boeing-Wichita, Major Gust Askounist and I as pilots had flown every hook-up that had been made using the hose method of refueling, starting with OPERATION DRIP through the refinement of the British system, the cross-over hook-up method, and, finally, the new American hose that was responsible for higher speed during the linkage, making controllability much improved. Unscheduled disconnects usually meant a broken hose, but with the stronger American hose it was rare that one would break up under any circumstances. And with conditions of rough air and darkness, the chances were slim, indeed, to even make a hookup. Gust and I both agreed that the Air Force should abandon the British hose method and proceed as soon as possible with the American nose-to-tail boom system, invented by Gust and developed by Cliff Leisy.

Gust and I both had an opportunity to hammer our point home after a demonstration of the British hose system to Colonel Charles Lindbergh. Lindbergh worked directly for General K. B. Wolfe, under General Hoyt S. Vandenberg, Chief of Staff of the U.S. Air Force. General Vandenberg had used Lindbergh at the end of World War II to look into air refueling, which the British already were doing.

I had first met Colonel Lindbergh when his son was kidnapped, though I knew he did not remember. I was part of the military guard at his home, from Fort Monmouth, New Jersey. I had also listened to his lectures at the Institute of Aeronautical Sciences at a special meeting in New York when I had been president of the Wichita chapter.

Lindbergh wanted to observe from the B-29 tanker because the services were having problems with in-flight refueling. The logical place was the navigator's position where he could watch the cross-over, the contact of the weight line to the drogue line and, finally, the latching of the two lines together as the tanker slid over to take its position aft and above the receiver. It was suggested that he move through the aft upper blister to observe the hose coming out of the tanker and into the receiver. In general, he agreed except for the journey with the chest-pack parachute through the tunnel. He decided to waive the tunnel trip for his first flight. However, he thought he might want to make another flight to ride the aft section to observe the hose reel operation, etc.

In late April 1948, early one morning, both tanker and receiver crews moved to the locker room to suit up. I had briefed Bob Williams on the extra-long flight suit and new chute harness for Lindbergh. As it turned out, the Colonel fitted the chute harness over his beautiful gray flannel suit, then, wearing his soft felt hat, climbed the boarding ladder of the front compartment. He settled in the navigator's front seat and put his headphones over the hat. I have many times wished that I had a picture of him during that flight.

Gust did a beautiful job of making the cross-over — the hookup between the B-29 tanker and the B-50 receiver. We had the hose in the receiver and fuel flowing in minutes. The weather was perfect. For test-flying we had a special air traffic control set-up with our own radio frequency for security, and on that frequency we had nothing to do with civilian traffic or the FAA. We operated out of the Old California Section of Wichita Municipal, now McConnell AFB.

I had noted that we were over Coffeyville, Kansas, at 8,000 feet. I pushed the mike button to tell Gust, when all hell broke loose. Number four engine ran away. I hit the feathering button and pulled the throttle. The right side aft observer hollered over the intercom, "Fire! Number four!" Glen Chambers, our flight engineer, advised that we were losing engine oil at a rapid rate. Every time I pushed the button to try to feather the propeller and reduce the rpm, more oil would leak and feed the fire. Glen noted that each time we pushed the button, we were losing about ten gallons of oil. This meant that we had eight to ten minutes to land. Possibly, the wing would burn off first!

Gust was at a disadvantage. He didn't know the territory. I advised him that we were over Coffeyville, so he put the nose gear down, and ordered the aft exit door open. I asked him with hand signals, pointing to the exit hatch, "Do we jump?" He advised us all over the intercom that we were slowed for one: "Take your choice."

I heard Lindbergh ask Gust, "What is your choice, Major?"

Gust replied, "I'm staying." He advised Lindbergh that it was his own choice to jump or stay. Lindbergh decided to stay. Gust landed the bird about 300 feet long on the north end of the old Coffeyville air base. In an orderly manner, the whole crew left the airplane, then assembled about 300 feet away, up wind. The bird's right wing was burning, but for some reason didn't explode. While we were standing waiting for the impact, Colonel Lindbergh made a remark. I will never forget my reaction when he said, very quietly, almost prayerfully, "We made the right decision, not to jump."

Gust's reaction was *mad,* quick! My reaction was the same until I thought, "We *did* poll everyone and actually made the decision collectively." Pilots, of course, were very sensitive to the decision-making process because the man who made the decision was the man who in the end would be either damned or praised.

I had advised Captain Freed, who had commanded the B-50 receiver, to pick us up as soon as possible and to tell my secretary, Flo McVey, that we were going to have the flight conference as soon as we returned. Captain Stultz arrived with the C-46 to get us. I'll have to admit, I was glad to be back in my office. Flo told me that Senior Vice President and Manager Earl Schaefer and Chief Engineer Harold Zipp had been advised of our problem by the flight-test tower radio operator. She suggested that I call Harold and tell him that we were okay, but that the airplane was ready for Oklahoma City's "Tinker Field Salvage." Harold remarked, "I'm going to Earl's office. When your conference is ready, call his phone and turn on the speaker." Cliff Leisy and Captain Freed had arrived, so we all were assembled. It was Cliff's project — he was the engineering brains of the whole in-flight refueling system, and Freed needed to be there as Air Force plant representative.

I made the call to Mr. Schaefer and turned on the speaker. At first, Schaefer was a little lost for words, I think because of Lindbergh's presence. Finally, he said, "Excuse me . . . I just want to tell all of you that you did a wonderful job."

Lindbergh shook his head in agreement and remarked, "Earl, I agree. A magnificent piece of airmanship on Gust's and E.H.'s part."

Our tower radio operator remarked that he had the recording tape on all through the emergency. This whole demonstration was a good example of how brittle and vulnerable the system could be under the best of conditions.

Gust stood up. "Colonel Lindbergh, I consider myself above average as a pilot. I think E. H. will agree that this hose method of refueling is not satisfactory in many operating conditions. For instance, at night in rough air, it is impossible to hold contact. The pilot's visibility is so poor that in good conditions we have trouble." At this point, I asked Flo to get a copy of my 21 April 1948 Wright Field report regarding flight testing various positions for the tanker and receiver during refueling. I asked her to read the last paragraph dealing with Gust's idea for the nose-to-tail tanker in front, receiver back and below. Lindbergh was impressed. I asked him if he would like to see a demonstration. He agreed, so I scheduled the flights at 9:00 the next day.

The weather was great. The air was smooth. Lindbergh chose the navigator's position in the photo airplane I was scheduled to fly. I wanted him to watch the actual flying from a distance to get the proper perspective. I told Gust, who was flying the tanker, "I'll come in close on your side. Let's sell this deal." If we could convince Colonel Lindbergh of our proposed method, he would recommend it to General Wolfe.

Bill Dwyer was my photographer in the rear blisters, and my new pilot, Doug Heimburger, was my copilot. Glen Chambers was the flight engineer, and Bob Williams and Dale Allison were rear observers. Gust and I agreed that we should form up southwest of the city over open country northeast of Medicine Lodge, Kansas. The other airplanes followed me.

Within 30 minutes, Gust was putting on his refueling demonstration like only he could do. Lindbergh remarked, "What a beautiful show!"

I answered, "Colonel, you're watching the greatest pilots of our time, except you. Sir, Gust invented this refueling position." He was dancing in and out of that rearward position putting on a wonderful performance for us all.

Lindbergh said that he had seen enough. On our way back, he thanked me for including him in my compliment. He added, "This position seems easy to fly. The visibility is excellent and you can see the movements of the front airplane clearly."

I commented, "Colonel Lindbergh, this is the way we should go in the future." I then called our flight-test tower and asked our operator to advise Flo to have Jim Wickham, our aerodynamicist, Verne Hudson, of preliminary design, and Cliff Leisy, the refueling design project engineer, present at our flight conference in one hour.

The flying-boom refueling method was established in that conference. Jim Wickham and Verne Hudson suggested that we use the Beech Bonanza "V" tail control system for the boom. Cliff Leisy said, "The flying boom." And with that, the system was named.

On 12 October 1948, Bob Williams had made the first nose-to-tail hookup. Cliff's design team had done a magnificent job in minimum time in working out this first flying boom. Gust was flying the receiver. Bob was later honored by the Air Force as "No. 1 Boomer."

The B-29 tankers and B-50 receivers that were being modified with the original British hose system in Wichita were delivered to the Air Force as contracted. Tanker flight wings were formed and history was made on long-range bombing using the hose system, even though it wasn't as good as the newer flying boom. Although production on additional hoses was stopped, the hose system continued to be used until the flying boom was installed in a sufficient number of tankers and receivers sometime in mid-1949. The most impressive hose refueling event to the world, and to the watching Russians in particular, had been the flight of the B-50 *Lucky Lady*. She had made a nonstop, 93-hour flight around the world, ending 6 February 1949.

The flying-boom development, under the able direction of Cliff Leisy, was transferred with him to Seattle from Wichita. The boom system brought on the Boeing KC-97 tanker and finally the KC-135 in its many stages. The eventual effect of the flying boom on the Air Force mission philosophy was profound; a high percentage of combat airplanes today are equipped for in-flight refueling using the boom system.

Those early lessons learned in flying the hose system made everyone involved try hard to find a better way, and Major Gust Askounist was first in line with his nose-to-tail procedure. I never felt that Gust received the proper credit for his initial idea — it had started in April 1948 with that flight report at Wright Field when he first suggested the nose-to-tail position. And

The first flying boom test. Bob Williams, in-flight refueling flight-test project engineer, made the first hookup on 12 October 1948. Bob was the “No. 1 Boom Pilot”; Major Gust Askounist, the Air Force test pilot, flew the receiver. (Photo, Boeing Airplane Company)

The flying boom hookup from the rear. Note the receiver offset to the left caused by the tanker slipstream. (Photo, Boeing Airplane Company)

the credit for developing the actual design for the flying boom must be given to Cliff Leisy and his team.

In the morning after the aborted hose refueling demonstration for Colonel Lindbergh, Mr. Schaefer had called me in and had asked if I would demonstrate the XL-15 liaison aircraft with Lindbergh aboard. I was pleased to have the opportunity. Schaefer asked about having Lindbergh actually fly the bird. My remark was, "Knowing the Colonel as I do, at this point he'll want me to show him. I'll ask him, however, to be sure. We'll have No. 2 XL-15 ready to fly by 1:30 today."

Schaefer said he'd let me know by lunchtime whether Colonel Lindbergh wanted to go. Flo later advised me at Plant One that Lindbergh would be at the bay window office at 1:30. He would have his chute; I would fly the plane.

Captain Freed and Lindbergh arrived on schedule. The Colonel again wore a well-pressed gray flannel suit. He was a sight with his gray hat and his parachute over his shoulder. Captain Freed said with a grin, "E.H., when you and the Colonel are finished playing, come back to my office." He waved. "Have fun."

Lindbergh smiled and, in his quiet way, remarked, "Is there something I should know about you, the airplane, or the flight?"

My answer was, "It would be a shame to miss what the little bird can do. I'll talk us through the procedures for the Eisenhower air show that I flew for the General at Fort Riley. I promise no low-level dangerous kid stuff. The little bird can do things that aren't normal for most other airplanes. The roll spoilers and the flaperons make some of these maneuvers possible. We'll do a max effort takeoff first. Zero flaps, brakes, engine full, release brake, run about 200 feet at still zero flaps, then tail up normal position for takeoff. At about 60 IAS, pull on full flaps, keep a little forward stick pressure, over an imaginary 50-foot obstacle. Nose should be slightly down; if not, a stall could cause the airplane to sink below the obstacle. With the very large flaperons and max lift with flaps 45 degrees down, the nose will be well below the horizon in level flight. On the other hand, with flaperons up to 15 degrees past neutral, the nose will be well above the horizon to get max lift."

Lindbergh remarked, "Interesting. Let's go through your show. It would be a waste of time for me to do any flying. This kind of flying takes practice."

I went through the whole Eisenhower show and talked through each maneuver. The Colonel got a kick out of it, particularly the square turn that pulled six Gs at 65 mph airspeed. When the "G" meter in the rear instrument panel registered 6, he said, "Impossible."

I continued on with attitude change, nose up to nose down, maintaining level flight, and, finally, a landing over an imaginary 50-foot obstacle, 200 feet to full stop.

Lindbergh remarked, "I see why the roll spoilers are necessary."

The Colonel and I went back to the bay window office. Flo was there, and we had a cup of tea all around, which brought out a remark from Lindbergh, "Just like jolly old England."

Our conversation then turned to the jet age that was coming up. I outlined our pilot crew training plans. First, General Electric's flight engineering course at the River Works at Lynn, Massachusetts. Second, pilot transition at Edwards Air Force Base on the twin-engine B-45, built by Douglas, and the F-80 Shooting Star, built by Lockheed. Third, Wright Field ejection-seat training and physicals. And fourth and final, actual flight time in the XB-47, the first heavy all-jet bomber that preceded the B-52, at the Boeing Test center, Moses Lake, Washington. I told the Colonel, "I plan to take all of the training myself and will take Doug Heimburger, the latest member of our pilot staff, with me on the first round of training."

Lindbergh agreed that I had a good plan and added that he would like to do about the same. He indicated he was aware that it would be a real challenge for older pilots. He smiled when he said this. I asked him, just before we left the office, if he would give me his autograph for my son John's scrapbook. I could see this hit a tender spot. He asked about John's age and school. I didn't tell him about doing the guard duty at his home in New Jersey during the kidnapping investigation.

I drove Colonel Lindbergh to Captain Freed's office and said goodbye at the door. His last words to me were, "E.H., you have a great group and I thank you for your time and special effort. I will pass on Gust's and your comments on the hose refueling. I hope to see you during your jet training."

I was sorry to see the Colonel leave. All I could say was, "Thanks. Give my regards to General Wolfe."

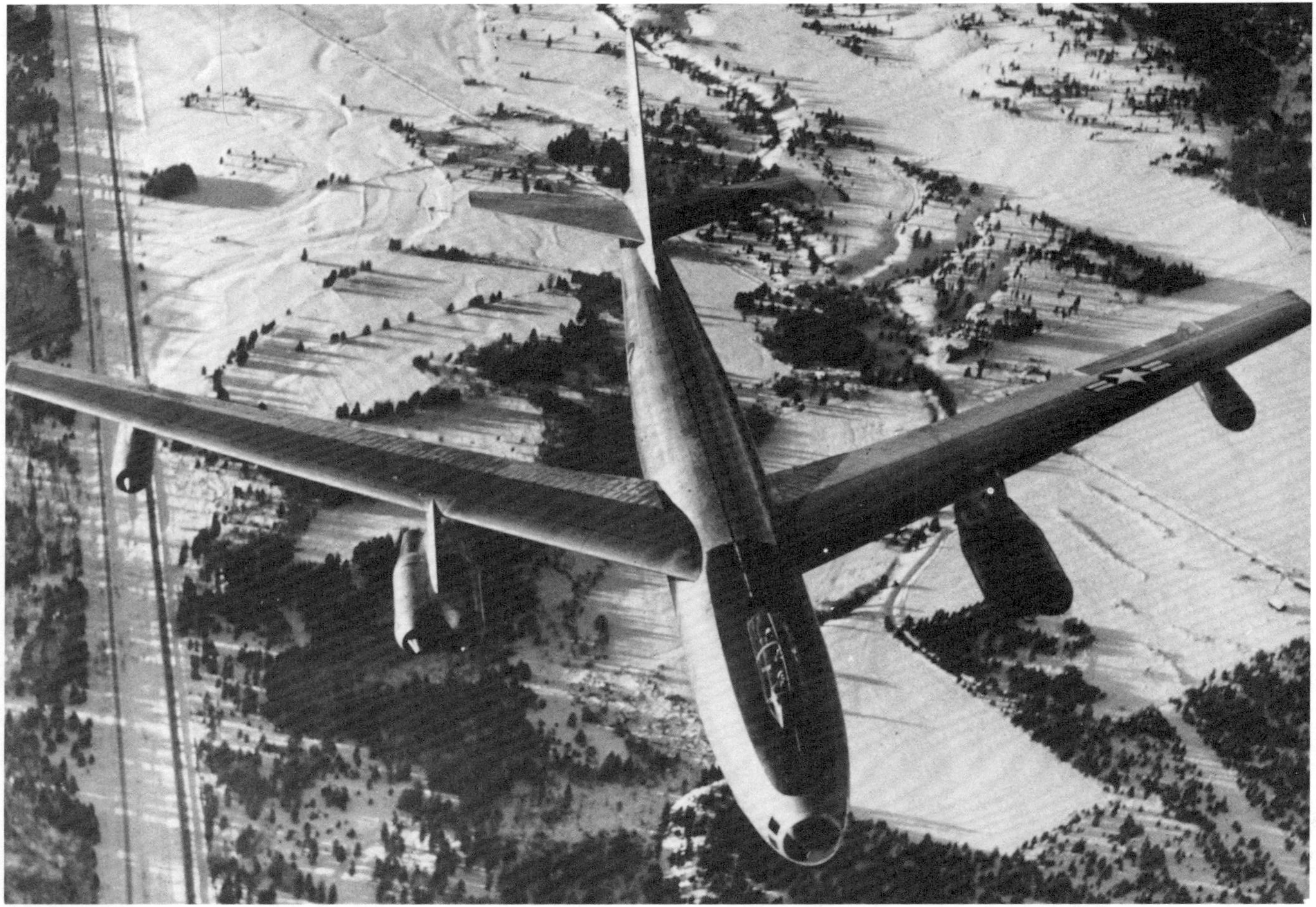

The XB-47, flown by Bob Robbins and Scott Osler. (Photo, Bob Robbins Collection)

The Jet Transition

The flight-test load had slowed after the refueling boom project had been transferred to Seattle. The engine test-stand work for the feeder line aircraft was completed. Bob Williams and Flo had finished the test documents. On 7 February 1949, Harold Zipp had called a meeting to lay out our future engineering work in Wichita. To the surprise of all, we got the word that the B-47 Stratojet project was going to be transferred from Seattle to Wichita. In addition, there would be some top management changes that would be announced in a few days. Later, Harold's final remarks were, "I will talk to all of you individually soon." He asked me to stay for my review.

Both Harold and Jack Clark, the assistant chief engineer, congratulated me that the flight-test department would stay under my control and that I should start hiring the necessary pilots and other staff to cover the total test program. Harold commented, "This is a serious challenge. The program is a big one and it is working in new ground, high altitude, Mach eight plus, with engines that are not good at best. Finally, E.H., Jack will take over the Mod Center in Tucson and I will be Wellwood Beall's executive assistant in Seattle." Beall was vice president and chief engineer for Seattle and Wichita. Harold continued, "This will take place as soon as your friend N. D. Showalter can get moved in as chief engineer. Mel Vanik will replace Chief of Staff Cecil Barlow, who is retiring. Bob Robbins will continue as chief B-47 project pilot until the program's Phase One (contractor's flight testing) is completed. I also suggest that you spend a few days with Bob and Scott Osler at Moses Lake. Scott is running landing tests for anti-skid brakes. This will give you a chance to make several landings without extra time. You're aware that the bicycle gear and outrigger wheels, as Bob says, make ground handling and landings very different from the B-29 or B-50." The wheels on the B-47 are in the center, with outriggers to keep it upright, instead of leaning on turns; the B-29 and B-50 had the standard tricycle gear with nosewheel and two main wheels.

My final comment to Harold and Jack was, "Jean and I will miss you guys." This was a fact, although N.D. and Bernice Showalter were also good friends of ours. The change, however, was a shock.

I called Bob Robbins in Seattle to set up a schedule to do some flying and get checked out on the No. 2 XB-47. He was receptive to the idea and noted that Scott Osler was well into the auto-braking test and that it would be wise to come as soon as possible. Jean and I flew to Seattle on Monday, 14 March 1949. We rented a car and were warned about snow in the Snoqualmie Pass, but we found the roads clear, though up to 20-foot cuts had been plowed through the drifts. It was a new experience for us. We found a small motel and restaurant in Moses Lake, the middle of the largest potato-farming area in the U.S.

The old air base was Boeing's test center. I spent Tuesday and Wednesday in flight conferences and reading the pilot's handbook. Thursday and Friday I spent flying from both the rear and the front seats. Scott also spent a lot of his time walking around the bird briefing me on every detail.

I remember on the last day, Friday, 18 March, I was in the front seat firing up the engines. I closed the canopy just before taxiing out for takeoff. The canopy latch indicator was not totally in the green; red was showing slightly.

Scott replied to my concern, "It's okay. We won't be pressurizing on this flight."

The flight was fine. Everything worked okay, though landings were hard for me. The flat approach and very little flare wasn't natural. The ideal attitude was for the aircraft's nose trucks to be off about one foot. If the nose was too high on touchdown, the nose trucks would slam down on the runway. As Scott remarked, "Your uppers will drop down on your lowers."

On Saturday, 19 March, Jean and I left Seattle at 7:00 a.m. for a stop in Los Angeles. I had scheduled an appointment that day with Dr. Lombard at UCLA to check on the Air Force safety helmet program. This was to be a great experience. The doctor reviewed the test at Wright Field, using "Little Oscar," the pet monkey, for impact studies. He told me, "Be sure to check on Oscar to let him know that you appreciate his work. He's a great little test pilot who needs friends."

I spent the whole day having my head form made from soft plastic material; the doctor promised my helmet within one week. I made arrangements for Doug Heimburger's helmet setting, plus three others. Doug and the others were the B-47 test pilots.

Jean and I stayed over at the Hollywood Roosevelt Hotel on Hollywood Boulevard and that evening went to Trader Vic's. I introduced her to the famous "test pilot drink" invented by Captain Jack Ridley, Chuck Yeager's right-hand man. I believe that Alvin "Tex" Johnston, later to become a renowned test pilot for Boeing, had a future hand in improving the drink. We enjoyed the old hotel, a favorite stopover for Boeing and many other aviation people.

We left Los Angeles at 10:30 a.m. on Sunday, 20 March, and arrived back in Wichita at 3:00 p.m. local time. On landing, I called Flo to let her know we were in. Her first words were, "Mr. Showalter called." She hesitated and with a strained voice added, "Scott Osler has been killed in Number 2 XB-47. The canopy came loose at the rear latch, slid back, and caught his head. Jim Fraiser, the copilot, brought the airplane in safely. John Fonesero was aboard." John Fonesero also was a pilot hired for the B-47 program.

I had been in that seat in that same airplane just two days before with Scott and Jim.

Flo went on, "Mr. Showalter wants you to call him at home at your convenience." She then added, "Why don't you go home? Everything is okay here."

That evening, I called N.D. He went over the accident in detail and assured me that the malfunctioning canopy locking mechanism that I had observed and reported earlier was being reviewed for redesign — in particular, the lock indicator.

"This will be done before either airplane flies again. John Fonesero will work with Jim Fraiser for the time being to continue the test program. As soon as you're ready, Number One at Moses Lake is scheduled to go to Wichita with your crew. I assume Doug Heimburger will be included." I assured him that Doug was in my plans, along with another pilot under consideration.

Showalter continued, "Bob Stanley, from Bell Aircraft, has given me three names: Tex Johnston, Bob Gorrill, and Ed Hensley, all experienced test pilots. Tex is very well known because of his 1946 Thompson Trophy win at the Cleveland air races. He's working for St. Louis Helicopter Company, based at Lambert Field, Missouri. I have his phone number and plan to call him the first of the week. I'm sure he knows Gorrill and Hensley." N.D.'s final remarks were, "Well, E.H., this accident is a very bad setback, but we have to carry on. Don't let this incident sour your feelings toward the B-47 program."

I assured him that it wouldn't, but I did stress the need for extreme care taken on all mechanical aspects of the test birds. I thought of Glen Chambers and his crew, who had maintained our B-29 test aircraft, *Sweet Sixteen.* We had test-flown that aircraft for 28 months without a scratch on either the airplane or a man. But the B-47, in contrast to the B-29, was a complicated bird and it would take even greater attention. N.D. agreed and asked that I brief Earl Schaefer and Harold Zipp. "Tell Jean that Bernice is looking forward to her move to Wichita, but is going to miss the sailboat. Let's keep in touch. Keep up the good work, E.H."

I called Tex Johnston and told him that Bob Stanley had recommended him for the senior project pilot's position in Wichita on the B-47. "Before you ask, Tex, this has nothing to do with Scott Osler's death. I'm putting together a totally new flight test-group. Your job is Number One." He was to be the chief project test pilot.

Tex asked, "Have you flown the bird?" I told him I had, and that the landing gear and touchdown angle was one objection, though the engine acceleration time probably was the worst. "Fifteen seconds from eighty percent power to one hundred percent is a long time if you are short on approach." The pilots used to say you could bore a hole in the ground and walk out before you got the power back! "The bird is a real challenge. The moving expenses and salary will start as soon as you agree to take the job. I know we can settle on a salary." Tex agreed to talk to his wife and call me the next day.

It was a mind-boggling problem to start an all new flight-test organization. We were beginning a new age in the flight testing of airplanes. Very little was known about test procedures for pure jets. General Electric was building the J-47 single-spool jet engine for the North American F-86 Sabre and B-45 Tornado, as well as for the Boeing B-47 Stratojet. And from ground testing and limited flight testing, a procedure had been developed breaking down the engine into test zones from the air entry to the end of the tail pipe. Dr. Robert Moss and Mr. A. M. Hatch of General Electric had organized a special school for this procedure at the Lynn, Massachusetts, River Works and had invited all flight-test people to attend free of charge. I made arrangements for Doug Heimburger, Jim Wickham, and myself to attend. The school was to be two weeks, eight hours a day, with evening study. Both Jim and Doug agreed that the program was a good opportunity. Jim knew Dr. Moss, who had taught thermodynamics for MIT; Jim had passed one of his courses. "E.H.," he said, "you'd better get ready for mathematical derivations all around the room, on every blackboard."

I told Jim he could help Doug and me if we got in trouble, and he just grinned and remarked, "Doug is the last one in the group to get out of school. He probably could help us both."

Doug said quietly, "I would be flattered."

Tex Johnston called as scheduled. He agreed to come to Wichita in a few days to take the company's physical and settle the salary and insurance allowed for the chief project pilot's job. I agreed to check on housing before he would arrive. Tex also helped me find the other test pilots, Bob Gorrill in Lawrence, Kansas, and Ed Hensley in Blackwell, Oklahoma. By the time Tex arrived, I had located for him a very nice three-bedroom place on East Douglas Avenue, four blocks from my home. N. D. Showalter and I had no trouble with the pay plan and insurance for Tex; both he and Doug Heimburger seemed to be compatible.

Tex privately suggested that I assign Doug to production test; he thought this would give Doug front seat and a lot of flying time. I could see that Tex wanted Gorrill and Hensley on his team. I agreed as long as there was no favoritism shown; we all had to work as a team. Tex agreed, but pointed out that he had worked closely with Bob and Ed and wanted them to feel comfortable with their new jobs.

I didn't press the point; however, I told Tex that he and Doug were scheduled to bring No. 1 XB-47 to Wichita as soon as Bob Robbins was finished with Phase One. Tex remarked, "We should both get checked out at Moses Lake." I told him that Doug would be back from jet engine school about 21 June, which would give Tex and his wife a chance to settle in their new place.

Jim Wickham, Doug Heimburger, and I arrived in Lynn, Massachusetts, on 3 June 1949, for the school. It was a pleasant experience. What little spare time we had was used traveling up and down the East Coast seeing all the historical sites from Gloucester to Boston Harbor, including some Boston night life and good food. Dr. Moss and Mr. Hatch topped our stay off with a graduation for the whole class of 35 held at General Electric's beautiful employees club on Nahant Point, Massachusetts. The good doctor roasted us all. He picked on Jim and me, in particular: "Here are two chiefs — one of aerodynamics and one of flight test. All through the course, they both appeared to be sleeping. I asked myself, how did they come out at the top of this class? Maybe that is why they are chiefs."

The whole crowd hollered, "They had their Indians do the work!"

The school was a great experience for us all.

Before we returned to Wichita, we stopped at the Wright Field Test Center. One of the final training requirements for the B-47 flight personnel was the ejection seat. All pilots and navigators were to be physically evaluated — in particular, our backs. All who passed could take the actual seat ride. I made arrangements for Tex Johnston, Bob Gorrill, Ed Hensley, Doug Heimburger, and myself to go through the program. Bob Gorrill was unable to meet the requirements. Tex, Ed, Doug, and I all passed the physicals and took the 90-foot rocket-propelled seat ride. It was, as Tex put it, "The biggest kick in the butt in history."

The fun was getting suited up and ready to go. The man in charge of our training was a salty parachute test pilot. He had jumped every new parachute and ejection seat rig that the government had purchased in the last 20 years. He showed movies of all of his seat jumps, from the cockpit and from his entry into the slipstream. The film in all cases continued on to show the parachute deployment and the actual seat separation from his body. The big thriller was an actual ejection from an airplane traveling 200 mph down the runway. Sergeant Wilson admitted that this was done several times with a dummy first. He also admitted that there wasn't much margin between busting his butt and landing safely.

The actual ejection seat ride took place in the large structural test hangar. The vertical test track was over 100 feet tall. The ride was about 90 feet. I remember drawing the short straw, so I made the first hop. The Sergeant sounded the warning horn. Every worker in the hangar crowded around to watch another poor soul trying to appear cool. I looked around at all those grinning faces and decided to have some fun. I knew a good scream at the top and words like, "Get me down, I'm hurt" would wipe some of those grins off.

The Sergeant yelled, "Are you ready?" I nodded my head and pulled the lever. The next thing I remember was that I was sitting 90 feet in the air looking down on the crowd. I forgot to scream.

When I was back down, Wilson and the doctor checked me out as okay. My crew wanted to know how it was. My answer, "You'll soon find out!"

They all went through the jump with flying colors. This was the last of the special training; they were now all ready for their flight checks in the B-47 at Moses Lake.

But first, Tex and Doug went to the Wright Field lab for their Lombard helmet fitting, and I went to see Little Oscar, who was back from the lab at Edwards; he was quite a character.

The B-47 and Pete Jensen

In 1949, in addition to everything else, the Boeing-Wichita plant went into production on the B-47A Stratojet in record time. However, neither engineering nor production anticipated the problems with the GEJ-47 engines, the "K" bombing system, and the radar-controlled tail gun. Each of these took special attention, extra man hours on the flight line, and many hours of repeat production flight time. We had some problems with the B-47, but this bird was well ahead of its predecessor, the B-17, especially when you think that the -17, only ten years earlier, had been an 180 mph piston-powered aircraft with a straight wing.

The B-47 was a radical swept-wing bomber whose 600 mph speed made it faster than the first- and second-generation jet fighters and certainly far superior to the piston-engined fighters and bombers of World War II, like the contemporary B-50 piston-engined bomber. The B-47's flight characteristics were very different from the piston-engined aircraft and its speed gave it different stall characteristics. But it suffered from the slow acceleration of the new jet engines, and its landing bicycle-gear configuration was horrendous. So it was not surprising that we had many new problems with it that no one had encountered.

The U.S. Air Force was gung-ho, of course, to fly the B-47 at its high cruising speed at high altitude — and that's what it had been designed for. But then the Air Force decided that the best approach for getting under radar was to fly down on the deck and toss the bombs out while in a tight climbing turn and heading onto the reciprocal course — what's called an Immelman. The result was that when you took a delicate high-altitude aircraft and thundered along at low altitude, the plane's structure was close to being overstressed. The aircraft had frequent structural failure above 5 Gs — the skin would loosen between the box spars in the thin wing. In addition to possible loss of life, this was an enormous maintenance problem — rivets had to be replaced with bolts — a problem way beyond then-current knowledge. This problem, however, was "cured" by a thicker wing on the subsequent B-52 Flying Fortress, introduced later, in 1952.

In the midst of all these snags with the B-47, the USAF was not at all happy when we told them that if they wanted to fly low and follow the terrain, they would have to limit the indicated airspeed to 410 mph. Boeing, the world leader in large aircraft structures, had never faced such engineering problems before.

The B-47 engines were actually our greatest concern: exhaust gas thermocouple failures on the ground and in flight, engine overheating during engine acceleration, and compressor stalls, causing turbine wheel damage. The flight line was becoming crowded with birds with problems of one kind or another. I finally recommended that flying be stopped — all B-47s grounded — until each airplane was thoroughly re-evaluated and ground tested both as a matter of safety and because we were eating up the test-flying budget with the extensive flying time. In mid-May 1949 I wrote a memo stating that we had to have better-trained ground service people to prevent flight-test duplications.

N. D. Showalter, our chief engineer, agreed with my grounding order. He reported the decision to Captain Freed, the Air Force plant representative. N.D. also called General Wolfe, head of Procurement at Wright Field, who also agreed. He advised N.D. that he and Pete Jensen, his adviser, would come to Wichita to review the problems firsthand. When I heard the name Pete Jensen, I told N.D. that I knew him when he was plant manager at Curtiss-Buffalo. N.D. remarked, "This could be good or bad. Is he good to work with?" I assured him that Pete was tough, but fair. He was not a pilot, but he was very experienced in all phases of aircraft manufacturing.

I will never forget the day the General and Pete arrived, 18 June. I was part of the welcoming committee in the Air Force flight office, as far back as I could get in the crowd. Pete, being over six feet tall, looked over everyone. He saw me, and said to the General, "Here is our problem." He walked over and put his arm around me. "General, here is the man who grounded the birds. He should be the man to fix them." This was in good fun, but some present didn't like it! Pete added, "I brought up this young man in the business. His real name is 'Cactus.' " Earl Schaefer, our vice president and head of the Wichita division, and the General then joined the fun. I knew, however, that I was stuck in the front of the problem.

The next morning I was called to Lyle Freed's office. Pete and the

General were there; to my surprise, no other Boeing people were present. However, the General put me at ease by explaining that Pete was going to stay on to direct the Air Force's interest in what he called a reorganization. He went on to ask if I would work with Pete as the Boeing representative on what he named the "Ramp Committee" — the work would be done "on the ramp," where all 125 aircraft were parked. "We want you to name the people — engineering and other technical people — who could handle the re-testing on the ground and flying the re-worked airplanes." This Ramp Committee was to be made up of the brightest and the best.

I agreed to work for Pete if it was okay with N.D. and Mr. Schaefer. I told the General, "There is nothing here that a little ground crew education and a few engine changes won't cure. And we will need to have General Electric cooperate." I knew that Lieutenant Colonel Bob Kuhn, B-47 project engineer, and Captain Jack Shaffer, in charge of engine procurement, would. The Air Force engine lab would need a little prodding to get some of the changes approved. "But," I added, "General, with all due respect, they, with G. E., are sitting on their hands. It will take you and Pete to get them to cooperate. We need an eight-stage bleed valve to prevent compressor stalling, a redesigned fuel control unit, and new EGT [Exhaust-Gas-Temperature] thermocouples that won't fail every flight. The K bombing system [an electronic one that had replaced the old Norden bombing system] needs a very precise 400-cycle current to do a good job on the bombing range. The present General Electric air-driven alternator system will never make the cycle accuracy requirements of three-tenths of a percent. The Sundstrand Company in Rockford, Illinois, has a mechanical transmission that does work, and Bob Kuhn knows about the development, but he can't get through the lobby field. We have some of that here, but that can be fixed, if N.D. and Earl Schaefer agree."

General Wolfe and Pete thanked me for a good briefing and the General added, "Be in a position to write up what you have told us. Don't be afraid to name companies and people."

The next day, 20 June, N. D. Showalter called me to his office to tell me that I had been selected to be Boeing's chairman of the newly formed Ramp Committee: "This will require you to give up your position as chief of flight. It will, however, improve your status. The company needs you to take over the power plant, equipment, and outside contract engineering. This includes Boeing's contact with the G.E. engine plant in Cincinnati. You'll have complete cooperation from Earl and me. The rest of the story is that you'll be made assistant chief project engineer under Tory Gamlin. Your salary will be increased substantially, including a cash bonus and stock participation. You should transfer Flo McVey to you and Tory. You'll need her help."

Lieutenant Colonel Bob Kuhn, Air Force engineering project manager on the B-47, Wright Field, Dayton, OH. Bob and the author shared successes and failures associated with the development of the first jet bomber.

N.D. finally remarked that the Ramp Committee would be terminated as soon as we got back on our delivery schedule. He went on to say, "I'd like you to put your ideas and personnel requirements in a memo to me and a copy to Earl Schaefer. I'd like to have this information before any action is taken."

My answer to all this was, "I wouldn't have it any other way." I could see, however, that N.D. was a little upset about General Wolfe's and Pete's invasion. This could involve a lot of engineering changes. N.D. was conservative and deliberate and did not want to move too fast. He was very careful about everything he did.

The Ramp Committee turned out to be a terrific group. Many at Boeing had white faces when the Committee first sat down to hammer out the B-47's problems, but we basically all had to work together. Pete Jensen handled everyone's fears well, and we all got through this "inquisition" and problem-solving period without firing anyone. We all found that we had a lot to learn about the aircraft.

My time as Boeing's chairman of the B-47 Ramp Committee was nearly three months. Glen Chambers, my flight line supervisor; Norma Rossie, my special secretary; Lynn Whiteside, from personnel training and management; Dale Allison, engine technician; and Bob Mariner, a tech writer, made up the team. Pete Jensen was the Air Force representative.

We started with new functional test procedures as the airplane left the factory door — we checked function on everything. We retested all airplanes on the line that had not been delivered to the Air Force. Glen and Dale then made ground crew selections called "The Best of the Best," and new crew members were trained and assigned to a supervisor from one of the "Best" group. Sundstrand came up with something like the automatic transmission used on a car, which was wonderfully simple, and we put it on the front end accessary drive of the GE-47SC engine. It was a surprise to Wright Field, but it worked so well that the basic concept is still in use on many military aircraft today. Called the Sundstrand constant-speed transmission, to drive the 400-cycle alternator, this device was necessary for the K bombing system accuracy. Wright Field had spent a lot of time and money with G.E. to perfect and build an air-driven system (using air from the compressor to drive the alternator), but could not get it accurate enough for the K bombing system.

Within a month, the Ramp Committee had over six crews working on their own. The B-47 was on its way. Captain Freed started to okay deliveries. New and retrained technicians were coming out of Lynn Whiteside's group training in good numbers. And the K bombing system and tail-gun radar were being ground tested to simulate flight. Actual range testing was done at Edwards AFB, California, later by Doug Heimburger. Pete Jensen, General Wolfe's adviser, was pleased; he didn't fire a single man or woman, though he came close many times. My standard retort would always be: "Why lose the brains? He (or she) is bound to fit in someplace."

Our top management was also pleased. It was nice to be around Pete again. It brought back some fond memories. And after our work was finished, Pete went on as a special adviser to Bill Allen, Boeing's great president. Tory Gamlin went back to Seattle engineering, and Bob Robbins stopped flying and was transferred to Wichita as chief project engineer, in Tory's position. I had no problem with this as Bob and I understood each other. He had been a great help to me when I started on the B-29 program. Later, Bob and I had a long talk about safety in the flight department. He complimented me on my record during the B-29 and in-flight refueling program and reminded me that my record was perfect except for the Lindbergh deal in the B-29 hose tanker over Coffeyville, Kansas, over which I had had no control. And he remarked about the death of Scott Osler, which I could see was a sore subject.

I eventually left the Boeing-Wichita flight-test department on 16 January 1951 to become the assistant chief project engineer. Tex Johnston, Doug

Heimburger, Bob Gorrill, Ed Hensley, and George Hanna remained as pilots. Dean Cunningham, one of Ernie Allison's old B-29 production test pilots, agreed to come back about the time I left, and so Dean and George Hanna finished the production testing on B-50 receivers and B-29 hose tankers to complete the old contract and finally flew the KC-97 boom tanker for the B-47 and, later, the B-52 production program. Tex and his pilot group, under John Fonesero, went on to fly both engineering and production testing through the B-47. Tex, however, was transferred in late 1951 to Seattle to fly the first B-52. My good friend Tex had done a wonderful job as senior project pilot during the beginning of the B-47 program in 1949. He and his crew distinguished themselves by setting a cross-continent speed record of 3 hours and 46 minutes. This helped to set right some of the lingering doubts about the production B-47.

With the KC-97 boom in-flight refueling (IFR) through 1952, the B-47 became even more impressive with respect to range. The KC-97 tanker subsequently gave way to the Dash 80 (-80) design, the forerunner of the great 707 airliner and the KC-135 jet (boom) tanker, which offered the perfect combination.

Tex Johnston and Lieutenant Colonel Guy Townsend made the first YB-52 flight on 15 April 1952, when Tex was chief of flight at Boeing's Seattle home plant.

The B-47 Development

After the Boeing Ramp Committee for the B-47 had been disbanded, on 1 August 1949, Pete Jensen returned to Wright Field to his position with General Wolfe. Long before he left, however, he had briefed N. D. Showalter and me on one of the most pressing problems that the Air Force had — the formation of B-47 combat wings. He listed the items in order of priority:

1. Pilot and copilot jet training
2. Maintenance personnel
3. Engine improvement and improved thrust with water-alcohol injection
4. High-speed, high-altitude bomb release with new ballistic tables
5. Ejection-seat development and production phase-in
6. Improved lateral control at high speed
7. Improved engine acceleration time for better handling on landing approach
8. Improved release mechanism for exterior fuel tanks. (If one tank fails to drop, the pilot loses control on landing.)

Pete went on to say, "There are some studies being made to find a suitable air base for jet transition from the P-80 (T-33) Shooting Star fighter to the B-47. I am at liberty to say that Wichita Municipal Airport is being considered because of its proximity to the Boeing plant and the long heavy runways. This would mean eventual elimination of civilian aircraft operation, including airlines."

My reaction was, "This will require a new airport, probably on the west side where there is lots of expansion room."

Pete closed the briefing by asking N.D. to form up his comments and stated that he would be back in a few days.

I knew that Pete and N.D. had me in mind to head up at least part of the "dog and pony shows" required to help in engine maintenance, pilot training, and bomb dropping. As to the engine modifications, Orva Douglas, Boeing's power plant project engineer, and I had been working with Captain Jack Shaffer and Colonel Bob Kuhn. Most of the engine changes suggested by the Ramp Committee were in the works, thanks to Jack and Bob working with General Electric. And Lyle Kuhns was also a big help, as our new local General Electric representative. He could get the word to the right people in the Cincinnati G.E. plant.

When N.D. had called me in earlier to review Pete's priority list for combat wings, I had proposed Orva Douglas as our B-47 engine group leader, Doug Heimburger or Ed Hensley as pilot reps, and Dale Allison and Glen Chambers as maintenance reps. Glen or Dale could make the ground crew selection.

Glen Chambers actually selected the first B-47 ground crews for the wings, and he picked good ones, as he had been in the Boeing flight-test department and knew exactly what was needed. In the past, there had been a gap between test pilots and ground crews over a couple of sensitive points: (1) pilots hated the fact that ground crews would run the B-29 engines hot on the ground to get good mag checks, and (2) ground crews tended to write off or ignore the pilots' "squawks" as fussiness. The B-47 Ramp Committee had helped to solve these differences and had helped us all get back onto schedule, working together.

But the Air Force B-47 crews for the combat wings needed a lot of training on the new aircraft and especially on the engines and the K automatic bombing system. The whole thing was a big change. We all went back to the drawing board, and many revisions had to be made to the aircraft for bomb training. For example, Tex Johnston found that the flaps were not interconnected. Everyone had missed this because the two different drives, the port and the starboard, were only connected electrically and not mechanically. On one flight Tex had experienced a flap differential extension — flaps at different angles — in the air, and he exploded over the radio.

The B-47 was big and was a structural conundrum. Boeing was the aircraft manufacturer with the most experience until it came to the B-47, and then we all started out from square one. Boeing's Wellwood Beall was an advanced designer for his time, and we were far ahead of everyone else, but the problems we encountered with the B-47 were due to our own inabilities to understand and handle such new and magnificent designs.

By about March 1950 I told N.D., "We can set up Smoky Hill Air Force

Base in Kansas as a test run to help get our show together. Next, March Air Force Base, California, then Barksdale Air Force Base, Louisiana, and finally, McDill Air Force Base, Florida. This will take about four weeks, total. I'll set up the schedule with each base commander as soon as Pete and you are ready."

I continued, "The bombing and ballistics test is a separate problem. We should suggest the Muroc [now Edwards Air Force Base] for our base of operation. The old north bombing range is grid-marked. This would be a good place to test for the K system. I'll get with Glen and Dale to select the men for the maintenance team. We must also check security clearance. I'm sure it'll be 'secret' or higher. Heimburger would be a good tech leader for the group. Dale Allison and I will go out to set up transportation and quarters. This will be a little problem because we must live off base. For full maintenance, we should plan on 40 to 45 men. The flight crew and bombardier will be military. This will be tight security because of the new thermonuclear *Little Boy* [nuclear bomb]. General Wolfe should set up this caper through Colonel Al Boyd, chief of the AF flight test division, and I want to see the flight crew assignment.

"This calls for top people with sound B-47 flight time. To my knowledge, the Air Force has not yet dropped the *Little Boy* at six hundred mph and forty thousand feet. It could be dangerous. Doug and I will write the flight plan if necessary."

N. D. remarked, "E. H., this sounds like a good plan overall. Flo has a recording that I'll play back to K. B. and Pete over our conference hookup. We better get the whole deal in the works."

After I left N. D.'s office, I stopped in on John Lewis, our armament engineer. In his work list were the 1,500-gallon drop tanks. He explained that his group was in the process of redesigning the whole release mechanism to use explosive bolts instead of the mechanical latch that was prone to stick in cold conditions at altitude. John added that the rig would be ready to test in about ten days. He would call me.

Dale Allison, Glen Chambers, and Orva Douglas were given the assignment to work up the script for the engine and maintenance side of our show. Ed Hensley, John Fonesero, and I had to assure that the pilot's part would be a group briefing: no flying unless it was planned in advance and with B-47 flight instructors. I personally pushed this non-flying approach because of safety, and we could cover more pilots' questions in our briefing. To get the pilots to talk would be a problem, but John had a good idea: "Brief them first in front of their Commanders and ask for their questions at the end."

We were ready with the "traveling show" in less than a week. The bomb-dropping test was waiting for airplane and crew assignment. Pete called

N.D. to inform him the bombing test would start before the "road show." So Dale Allison and I left for Muroc on 13 March 1950 to inspect our facilities. As to the airplane and maintenance crews, Major Bob Cardanas was in charge and Captain Joe Wolfe was assigned to show me around the base and to help in making arrangements for quarters and mess. Captain Wolfe was frank: "We have great hangars and service facilities for airplanes, but next to nothing for people. We can feed your men okay, but I would advise off-base for housing."

I asked about Pancho Barnes's nearby dude ranch, already widely known in flying circles as the "Happy Bottom Riding Club." I knew her from a previous trip. We could rent the whole place, if necessary. Our entire crew could get breakfast and dinner at Pancho's and lunch at the base. I was told that we could use military buses for transportation.

It was getting late, so I suggested that Dale and I go back to the BOQ and we all meet at Pancho's at 6:00. Joe agreed.

Dale and I had been a long way since 5:00 a.m. Central Standard Time in Wichita — a two-hour time difference. He reminded me that it was 8:00 back home and added, "I hope it will be a short evening."

I assured him that he would meet a lot of interesting people, and after a good steak and a little libation, things would look a lot better. Captain Wolfe arrived on time, along with Jack Ridley and Bob Cardanas. We all went to the lounge area, and soon Pancho also arrived. She was glad to see me and put on her best for Dale. I took her to one side and asked her about the chances of renting her operation for two weeks, explaining there would be about 45 people, one man per room, breakfast and dinner, bar closed after 6:30. I had a feeling that she could use the business, particularly the motel part. I asked her for a letter to me and a copy to Earl Schaefer on cost and so forth. She tried to put out her classic brand of "salesmanship" and asked if Earl could afford it. My reaction, "Pancho, this is serious. We can go to Barstow or Lancaster. We need an answer as soon as possible."

Bob Cardanas had settled a crew problem that came up between Captain Lyle Freed in Wichita and the flight test section at Muroc: Lyle had no one with experience to make up his crew. Thus Captain Sellers and Joe Wolfe and a bombardier from SAC headquarters had been assigned. Doug Heimburger was assigned to manage Boeing's interest in the technical side of the bomb-drop experiment.

Our maintenance crew from Wichita arrived at Muroc at about 4:00 p.m. local time on 18 April 1950. Glen asked that I call a meeting with everyone involved as soon as his crew had settled in; he explained that some of his men were on their first long trip away from home and added that there should be some firm rules laid down.

"Glen, with these surroundings and our assignment priority, you're right," I said. Glen was fair, but very firm. He made it clear that anyone violating normal behavior rules would be terminated and sent home.

I went back to Wichita after the first flight. The armament people were approaching the drop very carefully. Even the particular Doug Heimburger was satisfied. Bob Cardanas was flying chase, to get photographs, with a T-33. The pictures showed the bomb to be unstable as it entered the air stream out of the bomb bay; it shook around a little bit. I reported this to N.D., and also that the armament people anticipated some bomb shackle modifications, mostly adjustment. This was fine with N. D., who remarked that it was okay to start the special training road show with our first stop at Smoky Hill Air Force Base, Salina, Kansas. However, General Wolfe wanted to go over the script.

I suggested that he fly out to Salina for a preview and sit in on the first show. This would put a lot of pressure on our actors, but it would be good, because the General could enter into the questioning. His interaction would serve to pump up the pilots' participation. N.D. suggested that we get started; however, he thought it best that we listen to a dry run. Orva and Ed did a great job. N.D. was impressed, and so was I.

General Wolfe was a big help, though the Air Force pilots' questions, in my opinion, showed the lack of training, which our team discussed privately. The General showed interest and stated that Wichita would be the ideal spot for a jet training base. He remarked: "E.H., why don't you put the word out to Earl and N.D."

It was on that day that "Tent City" — the old Wichita Municipal Airport with tents for barracks and, eventually, McConnell Air Force Base — was born. George Hart and Dean Case, the two big fixed-base operators there, and QB brothers of mine, were very upset and blamed me. They had to leave the airport as soon as the City Commission approved the move, and though the new Westside Airport was in partial operation within 12 months, George Hart never forgave me. He would not move to the new Westside facility.

I had planned to go with our team to March Field, California, and also look in on the bombing test at Muroc. I insisted that we all stay at the old Mission Inn in Riverside. I had had the pleasure of staying there during the war but found it even better without the military crowd. That weekend, we had a chance to privately talk over our problems with the B-47. Ed Hensley made a shocking prediction that the lateral control difficulty would prove to be a serious restriction on speed; his estimate was about 200 mph. I later called N.D. to ask him if he knew how critical it was.

N.D. said, "I've discussed this with George Schairer. The worst of the problem is that the wing will take a trim set if the ailerons are rapidly deflected to start a turn. In other words, the upper and lower wing skins are slipping. If this comes up, frankly tell them that the problem is being evaluated. We're not in a position to discuss the matter at this time."

The March Field B-47 demonstration went well. There was no mention of the lateral control problem. However, at Muroc, Captain Wolfe and Doug Heimburger privately asked me if I had heard that the bird was going to be placarded to a much lower top speed. I confessed that I knew about the problem, but didn't know about the speed restriction. I did know, however, that the problem was under study by Boeing. Joe Wolfe was very concerned because he had been assigned to do a performance test program on the B-47E model. He mentioned that he had noticed during the high-altitude, high-speed bomb test that after the roll-out from a high-speed turn, the inside wing of the turn would trim light and also exhibit adverse yaw during the turn. He went on to say, "I've written up the problem of lateral trim change at least twice. Our mechanics would realign the ailerons and tabs on the next flight. The lateral trim would be okay. Our mechanics confessed that they had not found the ailerons out of trim in either case."

Doug casually remarked, "Maybe the landing shook the wing back in trim."

This was a good point. A few weeks later, when the B-47 was restricted to 410 mph top speed, the fix was to add close tolerance drive bolts along the skin line of the front and rear spars for the total span. This was effective, but very costly, both in time and money. All B-47E models were corrected by production change or modification.

Doug Heimburger and I returned to Wichita after the bombing tests in early May 1950. I checked in with N.D., and to my surprise, he informed me that General Wolfe and General Joseph "Smokey" Caldara, the Air Force safety expert, wanted me to head the pilot, aerodynamics, and engine accident investigation team. Bill Arundale and Bob Sivley were assigned to the structural and mechanical team. N.D. explained that with the increasing flight hours being put on the B-47, there would be accidents of all kinds. Therefore, our best approach would be to analyze the causes and to recommend the proper corrections. I agreed, because Pete Jensen's eight original combat wing development problem items were under control, even to the pilot and aircraft service training, thanks to the new Wichita training base.

My first accident investigation was in mid-summer 1951. The airplane was based at March AFB, California. The accident had occurred over the mountains, southwest of Wendover, Utah. A bomber was being refueled by a KC-97 tanker when the bomber stalled and went into a spin. The pilot was unable to recover to avoid a mountain top at 9,500 feet above sea level. The airplane and crew were lost. This was a flight planning error. Thirty miles east of the mountain was a valley with a floor less than 3,000 feet above sea level. The tanker pilot should have insisted that the refueling take place over the valley, instead of the mountains. The bomber pilot could have recovered from the spin with the extra altitude.

On Labor Day weekend, 1951, the Wichita-Boeing plant was closed, except for the flight test department. Jean and I were living on North Pine Crest Avenue in Wichita at the time. Dale Allison and I were sitting out back with a couple of Scotches when I heard two B-47s coming close by at about 500 feet. Seconds later there was a terrific BOOM. Two flight crews headed by Doug Heimburger were on duty that day; Doug Heimburger and Steve Gatti made up the first and Chester Coltharp and Eric Oppenheimer the second. Both crews were out testing several airplanes that had to be flown before delivery to the Air Force.

From the radio tower operator's testimony, Chester Coltharp had asked the tower for a gear check in-flight. Doug apparently had heard the message and had asked Chester for his location. He said that he would do a fly-by to check the gear. That was the last message from either airplane. Apparently Doug had tried to fly under Chester to check out the gear malfunction and had collided with him. The airplanes went down about one and one-half miles northeast of the Beech plant. There were no survivors.

I asked to be relieved from any investigation duty. N. D. Showalter headed the investigation personally. The memorial rites were held at the Central Christian Church in Wichita, Wednesday, 5 September 1951. All four men were honored. This was a sad time for my family. We were especially close friends with Doug and his wife, Harriet, outside of work. It was also a time for me to reflect on my accident-free test pilot days. Later N.D. said, "E.H., you stopped at the right time." I agreed.

Conclusion

The B-47 was a milestone in my total engineering career in the aircraft business. I was the last project engineer, and thus stayed with the program until it was finished — I wound it up. I did all the final "miles and miles" of changes necessary for completion.

The B-47 was the biggest thing to happen to the aircraft industry as well. In the early 1950s, the B-47 was thought to be state-of-the-art. It was the first high-performance all-jet bomber proved to be operational at over 40,000 feet and to speeds over 600 mph. At one time it had held all the records for speed and distance for an airplane in its class. Boeing and the Air Force had learned a lot, from both the B-47's basic design and operation, and from its subsequent derivative models.

At about the same time, the Soviets had made great strides in their defense against high-altitude bombing. The Surface to Air Missiles (SAM) and the development of interceptor aircraft like the MiG series, particularly the MiG17 onward, with air-to-air weapons, ended the use of the B-47 as a high-altitude bomber.

The Air Force thus developed a different bombing technique called "toss bombing." I had the pleasure of watching Boeing's top test pilot, Dick Taylor, demonstrate this method around 1954. He would approach the range target at about 100 feet above the ground and at max speed, pull the bird up into an Immelman (half-loop), then roll out on top. The idea was to toss the bomb at the 90-degree point in the pull-up. This maneuver would put the airplane safely out of the bomb blast. However, the B-47 airframe was not designed for this type of high G forces, and the airplane structure suffered to such an extent that several aircraft and crews were lost.

There also were other problems, of course, that caused serious accidents. One that comes to mind was the lateral control hydraulic boost system. If there was a partial hydraulic pump failure on the main system, the shuttle valve that normally shifted the system to the emergency pump would stick in dead center, locking the aileron controls — in some cases, in a deflected position. Several fatal accidents were caused by this type of failure. I was convinced that my friend Captain Joe Wolfe in 1949 had been killed in a B-47 because of a control malfunction.

In the early fifties there was also exciting news — the B-52D heavy bomber production would start in Wichita. B-47 production was thus suspended there in the mid-fifties, and most of our engineering personnel was transferred to the B-52 program. My friend, Jack Clark, returned from the Tucson modification center and was made project director of the B-52. It was a pleasure to have him back in Wichita. My old accident investigation team was still on line: Dale Allison, Glen Chambers, Ed Hensley, and Orva Douglas were assigned to the B-52, but were available to me as project engineers for the B-47 wind-down and for the B-47 accident investigation team, as needed. We worked many accidents from Canada in the winter to the desert in the summer. Bill Arundale was transferred to the Wichita B-52 program as project engineer on the new B-52 wing, and Bob Sivley continued in Bill's place.

A total of 2,000 B-47 bombers had been produced by Boeing, Lockheed, and Douglas before the B-47 was phased out in favor of the B-52. The last B-47 was produced by Lockheed in February 1957. On 12 February 1958, N. D. Showalter accepted my resignation with regret. N.D., in his quiet way, said, "E.H., you and I have come a long way since the B-29 days."

I had flown with some great pilots throughout the years. Dean Cunningham, starting in 1949, was the B-29 production flight-test pilot later involved with in-flight refueling production checks for the B-52. He was quick and precise in emergencies. Tex Johnston could fly like a bird and was also technically good. He would talk his way through any tests over the radio, to everyone's fascination.

Leonard Small had engineering degrees and had been an instructor pilot during World War II. He was the calmest young man I have ever seen, cool and collected, a stabilizer amongst the crews. He never had any personal or professional problems, and he could copilot, too. He never flew with Tex, as they were of slightly different generations, and they each had their own crews.

Jack Peacock and Jack Jones of the B-29 days were two other fine pilots — engineering types, excellent on the safety side.

And Doug Heimburger, of Springfield, Missouri — his father had been a

Elton Rowley, on his last day with Boeing-Wichita, 12 February 1958. (Photo, Boeing Airplane Company, by Bill Dwyer, Boeing photographer)

medical missionary in China, and he had himself gone to school there. Though he was an excellent engineer, and he was different, I hired him. He was very much like the Chinese pilot, Colonel Hun Chan Tang, whom I knew from the Chinese trainer program; you were never sure if he had heard you.

He was so damn good, you would tend to forget his faults. This was evidenced when we were testing the Boeing light XL-15 Scout — the radical development of the spotter aircraft used in World War II. Apart from its single boom after fuselage with twin tails, the XL-15's unique feature was the flaperons — barn-door flaps — which could go either 10 degrees up or 40 down. But in gusty conditions, lateral control could be poor.

One very windy day Doug and I were going to fly this little bird, and I told him that we did not use full flap on such a day. He took off and came back into land. On the second landing he applied full flaps, and when he touched down, the machine swung around and around in an exceptional ground loop. Like the Chinese pilot, he always wanted to find out for himself.

It was wonderful to know these people — a privilege — even as their boss. You were thankful that they were there. I never had a bad test pilot or crewman.

After I retired, I missed all the excitement and pressure of the industry for a while, but got back on the "hot seat" when I was made president of the family automobile business. Being a new kid on the block, the Wichita Chamber of Commerce asked me if I would take the job of president of the Chamber's Ambassador Club, the membership sales arm. In 1959-1960, my team sold more memberships than ever before in the Chamber's history, earning me a life membership. I found out much later that Earl Schaefer had recommended me for the job. Because of my experience with the military, hotel operator Walt Schimmel, the head of the Chamber's newly formed military affairs committee, asked me if I would be his co-chairman. This job lasted from 1961 through 1971.

Colonel Bob Cardanas became General Cardanas, Commander of McConnell Air Force Base in 1966. He was responsible for the F-105 Thunderchief training of the tactical Air Force, a program vital to the Vietnam War. In 1967 Bob formed a Chamber group called The Friends of McConnell (FOMs), a civilian group of ten men from the Wichita business community that smoothed the way for the military many times. Bob and I still keep in touch.

One of my fun flying projects in later years was the World War I Curtiss Jenny replica, which I built. My father-in-law, a Curtiss employee during World War I in the Buffalo, New York, plant, was a wing carpenter and wood

General Robert Cardanas, a close friend and military adviser — the one who insisted that the author write this book.

Jim Greenwood, aviation pioneer and senior aviation executive — also the author's friend and adviser.

NBC "Today Show" host Jim Hartz and the author, 27 September 1975. The occasion was Hart's television show featuring "The Kansas Story." The theme was "from Jenny to Jet." A Lear jet is in the background. (Photo, Jim Greenwood)

E. H. Rowley in 1978 at Barksdale AFB, LA, to fly his Curtiss JN4D Jenny in an Air Force Day air show. Dr. Arthur Metcalf, chairman, U.S. Strategic Institute, and General Ira C. Eaker attended the show.

artist, and a key man in the construction of my bird.

I flew the Jenny first in May 1969. This maiden flight was a great experience. What has been written about the Jenny is true: "She is no lady."

I flew the beautiful bird in many military and civilian air shows around the country, and she was a proud winner in every show. Then on 5 November 1985, our family donated "Jenny" to the Combat Air Museum in Topeka, Kansas. The flight from Wichita to Topeka was my last as a pilot in a 56-year aviation career. It is very gratifying to me that the Jenny is still in beautiful flying condition.

I hope that those who read my story will have a better understanding of aviation's history from the 1920s to 1960 — that "time before space." This period of aeronautical development was indeed beyond the industry's wildest dreams. And to have played a small part in it was certainly beyond mine, too.

Index

by Lori L. Daniel

1st Battalion Mechanized Cavalry, 64-65
6th Signal Corps, 54
15th Observation Squadron (OBS), 14-15, 22-24, 31-32, 34, 43, 51, 53, 108
20th Bomber Command, 159
61st Coast Artillery, 62
Twentieth Air Force, 159

— Aircraft —

707 airliner, xi, xv, 216
-80 (Dash 80), 216
AT-6 (North American), 103, 114
AT-15 (Boeing), 135
AT-18, 91
American Eagle, 73
B-17 Flying Fortress, 138, 144, 211
B-25, 191-192
B-29 Superfortress, x, 138-141, 144-145, 147, 150-151, 154-155, 158-159, 161-163, 165, 168, 174, 176-177, 179, 182-188, 191-192, 194, 196-197, 199, 206, 208, 215-216, 218, 225
  Sweet Sixteen, 147, 150-151, 155, 157-162, 208
B-32 (Convair), 144
B-45 Tornado, 203, 208
B-47 Stratojet (Boeing), x-xi, xv, 194, 207-210, 212-219, 221-225, 205-206
  combat wings, 217-218, 222
B-47A Stratojet, 211
B-47E, 222
B-50 Superfortress, 163, 168-169, 182, 187-188, 190-193, 197, 199, 206, 211, 216
  Lucky Lady, 199
B-52 Flying Fortress, xv, 203, 212, 216, 225
B-52D, 225
BT-13A (Consolidated Vultee), 165
Beech Bonanza, 46, 199
Beech Staggerwing, 110
Beech Travel Air 2000, 73, 81, 83-84
Bird, 73
Curtiss JN4D Jenny, 1, 227, 230-231
Curtiss C-6 Oriole, xi, 9-10
C-46 Commando, 138-139, 193-194, 197
CW-20, 95, 97
CW-22, 95-96, 101
DC-3 "Gooney Bird," ix, 122, 193
Empire flying-boat, 195
Executive 7W (Spartan), 103, 110, 112, 114, 118-119, 123-124, 130, 132, 134, 140-141
F-80 Shooting Star, 203
F-86 Sabre (North American), 208
F-100 Super Sabre, xv
F-105 Thunderchief, 227
Fokker, 42, 51
Ford Trimotor, 42, 51, 53
Greve class racer, x, 82
Handley Page Harrow, 177, 195
J-5 Waco, 48
KC-97 (Boeing), 199, 216, 222
KC-135, 199, 216
L-3 Cub, 174
L-6 Interstate, 163, 166, 168-169
L-15 Scout, 166, 170, 174
Lear jet, 229
Link Trainer, 149
MATS transport, 70
ME-109, 103
MiG, 224
MiG17, 224
N2S (Boeing Stearman), 110, 139
N3N (Boeing Navy), 119, 135
NP-1 (Navy trainer), x, 110, 112-115, 117-121, 123-124, 128-129, 131, 133, 135-137, 139, 142
  The Spirit of Spartan, 136-137
0-19 (Thomas Morse), 14-15, 27-28, 32, 34, 51, 53, 58, 108
O-47, 108
OPERATION DRIP, 178-179, 181-183, 187-188, 191, 196
  Gusher, 188
OX-5 (Curtiss), 72-73, 80
P-36 (French Hawk), 103-105
P-40, 102
P-51 Mustang, ix
P-80 (T-33) Shooting Star, 217
PT-3, 27-28, 32, 36, 62
PT-13, 172, 174
PT-13A, 167, 173
PT-13D, 165
PT (Boeing), 110, 139
Robin (Curtiss), 73
SBC-3, 90
SBC-4, x, 89-92, 94-95, 98, 105
Stearman, 103, 131, 135, 141
Stinson, 85
  Gullwing, 165
T-33, 221
Taylor Cub, 87
Turkey Hawk, 89, 94-99, 101, 103, 105
Waco 9, 73
Waco 10, 73
Waco Taperwing, 76
XB-29, 144
XB-47, 203-204
XL-15 Scout, 163, 168-171, 173-176, 202, 227
XP-55 Ascender, 139
YA18, 88
YB-52, 216
YL-15, 169
Zeus, 114

— A —

A&E (Airplane and Engine) Mechanic, 89
ADF (automatic direction finder), 54, 149
ADI (water injection test), 161
Adirondack Mountains, 2
Aerometeorograph, 20, 34, 39-40, 55, 59-61, 89, 96
AFPR (Air Force plant representative), 177
Air mail, 51-54
Alabama, 192
  Maxwell Field, 191, 193-194
    War College, 191, 193
Allen
  Bill, 215
  Eddie, xi, 97, 114, 138-139
  William M., 184
Allison
  Dale, 193-194, 198, 215, 218-220, 223, 225
  Ernie, 144, 155, 216
America, xi, 45, 53, 58
American, 180, 188, 196
Anciente and Secret Order, 46
Andari, General Alonis, 99, 101-102, 106
Anderson, Lieutenant, 54
Arizona
  Green Valley, xii
  Tucson, 205, 225
    Mod Center, 205

Armstrong, Neil, x
Arnold, General Hap, 46
Arundale, A. G. (Bill), 184-185, 222, 225
Askounist, Major Gust, 178-179, 182, 185, 188, 191, 194, 196-200, 203
Associated Press, 188
Austin
  Jean, 13, 43, 46-50, 62-63, 70-71, 75
    See also Rowley, Jean
  Louise, 47, 49-50, 77, 86, 127
  Ralph, 47, 49-50, 77-78, 86, 127

— B —

Balfour, Captain Max, xi, 103, 109-110, 124, 129-132, 134-136, 142
Barlow, Cecil, 205
Barnes, Pancho, 220
Barnett, Archie, 126, 128
Barnstorming, x, 1, 87
Barrett, Chief Master Sergeant, 62
Barrows, Arthur, 183
Baxter, Sergeant, 66-67
Beall, Wellwood, 205, 218
Beaman, Ann, 146-147, 158
Beam navigation system, 53-54
Beaver, Dr. Jim, 162
Becker, Sergeant, 165
Beckford, R. D., 157
Beech Aircraft, 137, 176, 223
Bell Aircraft, 87, 163, 207
Bell telephone system, 51
Benny, Jack, 75
Benson, Dick, 94
Berlin, Don, 88, 91, 106
Biers, Doris, 129, 131-132
Bishop, Master Sergeant Bill, 18, 42, 53, 55, 57, 59, 65, 69
Blaylock, Ray, 90-92
BMEP (brake mean effective pressure), 169
Boeing Airplane Company, x-xi, xiii, xv, 126, 137-143, 145, 179, 184-189, 193, 206-207, 212-215, 220, 222, 224-225, 227
  Ramp Committee, 213-215, 217-218
  Wichita, 110, 132, 139-140, 145-147, 151, 155, 158-159, 161-162, 164-165, 168-169, 175, 177, 183, 193, 196, 211, 215, 217-218, 223, 226
    Engineering Division, 183
    Plant I, 158, 164-165, 169, 176-177, 183, 202
    Plant II, 158, 164, 177, 183, 185, 187
  Seattle, 144, 163, 177, 191, 216
Boeing Lake, 171
Bombs
  Atomic, x-xi, 155
  Grand Slam, 154-156
  "K" bombing system, 211, 213, 215, 218-219
  Norden bombing system, 213
  Nuclear (*Little Boy*), 219
  Toss bombing, 224
Boston Bull, 17, 25
Bowman, Captain Jim, 25-26
Boyd, Colonel Al, 219
Braden, Jim, 71
Bradley, Colonel Jim, 191-194
Brann, E. C., 156-157
Britain, 66
British, 177, 179-180, 183-184, 187, 192, 195
  Royal Air Force, 177
Brodie gear, 169, 173
Bronson, Mike, 76
Brown, Lloyd, 114, 129-131, 134
Brownfield, Lieutenant, 23
Brown recorder, 159
Burr, Howard, 105

— C —

CAA (Civil Aviation Authority), ix, 11, 85, 109, 111, 115, 124, 157
Caldara, General Joseph "Smokey," 222
California, 128, 158
  Barstow, 220
  Edwards Air Force Base (Muroc), 203, 210, 215, 219-222
  Hollywood, ix
  Los Angeles, 146, 206-207
    Hollywood Boulevard, 207
      Hollywood Roosevelt Hotel, 207
    Trader Vic's, 207
  March Air Force Base, 219, 221-222
  Muroc, *see* Edwards Air Force Base
  Riverside, 221
    Mission Inn, 221
  San Francisco, 53, 195
Calley, Jim, 167
Cardanas, Colonel Bob, 179, 182, 220-221, 227-228
Case, Dean, 221
CAVU (clear and visibility unlimited), 20, 92
CG (center of gravity), 102, 114, 118
Chambers, Glen, 146-147, 151, 191-192, 197-198, 208, 214-215, 218-221, 225
Chance-Vought, 103, 119
Child, Lloyd, 88, 91-92, 94-97, 99, 101-102, 104-105
China, 102, 165, 168, 173, 175-176, 227
  Sea, 70
China-Burma-India theater, 159, 161
Chinese, 168, 172-173, 175, 227
  Nationalist, 165
Church, Bob, 85
Clark, Jack, 140, 144-146, 151, 182, 205-206, 225
Clark Y basic airfoil, 83
Claude, John, 79
Cleveland air-hydraulic landing gear struts, 44
Cleveland National Air Races, 46, 207
  Thompson Trophy, ix, 207
Cleveland, Pop, 44, 46
Code Word (CW), 52-53
Cold War, 168
Collins, Jimmy, ix
Coltharp, Chester, 223
Communists, 168
Consolidated, 87
Cornell, Billie, xiii, xvii
Craig, Lt. General Howard, 183
Crawford
  Dick, 57, 59, 69
  Jim, 96, 162
Crossfield, Scott, ix
Cunningham, Dean, 216, 225
Curtiss, Glenn, 1, 88
Curtiss-Wright, x, 87-89, 98, 115, 137, 139
  Buffalo, NY, 89, 91, 94-95, 101, 103-108, 110-111, 117, 123, 132, 138-140, 146, 151, 212, 227
  Milwaukee, WI, 70, 76, 80, 89

Curtiss-Wright (cont.)
New Jersey, 74
Airport Division, 74
St. Louis, MO, 94-95, 117

— D —

Dalley, Lieutenant Ben, 20-21, 24, 26, 28-29, 33, 36, 40-41, 44-46, 48, 50, 52, 58, 62, 69-70, 107
Dalrymple, Wayne, 151, 165
Davey
Bill, 88, 105
Jack, 88-92, 94-95, 97-99, 101-102, 105
Depression, 35, 75
DePussen, Sergeant Pierre "Poncho," 31-38, 40-44, 51-53, 58, 64, 69, 97, 102
Dixon, _____, 69
Doolittle, Jimmy, ix
Douglas, 128, 203, 225
Douglas
Donald, ix
Orva, 218-219, 221, 225
Drake, Hubert, 117, 139
Drum, Captain Joe, 147-148
Dunn, Lindsey A., xiii
Dwyer, Bill, 198

— E —

Eaker, General Ira C., 230
EGT (Exhaust-Gas-Temperature), 213
Eisenhower, General Dwight D., xi, 174-175, 202
Elliott, Bob, 165
Enderton, Otto, 9, 32, 48, 70, 84-85, 87-89, 102
Engines
3350 (Curtiss-Wright), 138, 144
GE-47SC, 215
GEJ-47, 211
Harley motorcycle, 7
Henderson motorcycle, 7
J-47 single-spool jet, 208
Lycoming, 165, 168, 173
Pratt & Whitney, 38, 168
Tank Model 63, 73-74, 81
Tank "Sky Motor," 73-74
Wright Cyclone, 159
England, 177, 183, 187, 203
Oxford University, 143
English, x, 16, 99, 173, 177
Everest, Pete, ix

— F —

FAA, 197
Farrell, Sergeant, 56-62, 69, 97
Fausel, Bob, 102, 105
Fendrich, Captain Nels, 145-146, 150
Finley, Staff Sergeant, 17-18, 20, 25, 28-29, 31, 34, 41, 55, 65, 69, 95
Fleck, Captain W. E., 175
Florida
McDill Air Force Base, 194, 219
St. Petersburg, 86
FOMs (The Friends of McConnell), 227
Fonesero, John, 207, 216, 219
Forrestal, Admiral, 126, 130-131, 144
Foulois, General Benny, 54
Fox Movietone News, 56-57, 65-67
Fraiser, Jim, 207
France, 103
Paris, xv, 8
France, Charles, 94-95, 97
Frankie, Elmer, 80-81, 85
Freed, Captain Lyle, 177, 179, 185, 191-193, 197-198, 202-203, 212, 215, 220
French, 103-105
Canadian, 31

— G —

Gannett newspapers, 9
Gamlin, Tory, 213, 215
Gardner, Major General Gran, 183
Gatti, Steve, 223
General Electric, 203, 208-209, 213, 215, 218
Gentry, Jay, 110, 114, 117-118, 124, 129, 131-132, 134-135, 142
Georgia
Atlanta, 131, 134, 139, 150
Naval Training Station, 132-133
German, 16, 70, 74, 99, 151
Air Service, 9
Getty, J. Paul, 125-132, 134-137, 139-140, 142-144, 146
Getty Oil, 126
GFP (government-furnished property), 122-123, 140
Gifford, Frank, 88-89, 91-92,101, 106
Glider, 1-2, 4-5, 77, 105, 107, 117, 126
Baby Bowlus, 117
R-100, 116, 139
Gorrill, Bob, 207, 209, 216
Greenwood, Jim, xii-xiii, 228
Greer, Jim, 165
Grey, Lieutenant Harvey, 94-97, 102, 108, 117, 138-138
Gubser, Gene, 124-125, 141-142
Gwinn, C. W., 103, 109-111, 124, 127

— H —

Halverson, Second Lieutenant, 17-18
Hamilton Standard Propeller Company, 72-73
Hanna
Dean, 216
George, 165, 168, 175, 192, 216
Harlow, Rex, 182
Harris
Colonel (later General), 145, 148-150
George, 151
Harrison, J. S., 157
Hart, George, 221
Hartz, Jim, 229
Hatch, A. M., 208-209
Hatfield, Colonel "Pappy," 191-194
Hawaii
Pearl Harbor, 124, 188
Hayes, Dick, 112-115, 117-119, 123-124
Heimburger
Doug, 198, 203, 207-210, 215-216, 218-223, 225, 227
Harriet, 223
Hensley, Ed, 207, 209-210, 216, 218-219, 221, 225
Hitler, Adolf, 103
Hoffman, Ray, 141, 144, 146
Holcomb
Clark, 1-8, 49-50
First Lieutenant, 19-20, 23, 42, 52, 69
George, 48-49
Rosmond, 49
Stella, 2-3
Holderman, Russell, 9

Hovaker, Paul, 117
Hudson, Verne, 173-174, 199
Hufgard, Paul, 105

— I —

IAS (Indicated Air Speed), 179, 202
IFR (in-flight refueling), 177-188, 191, 195-196, 198-203, 205, 215, 222-223, 225
  American nose-to-tail boom, 196, 199
  Boeing system, 195
  boom, 216
  British hose and grapnel, 195-196
  Flying boom, 195
  Probe-and-drogue, 195
In Flight Refueling, Ltd., 177, 183, 187
Institute of Aeronautical Sciences (Wichita, KS), 168, 196
Illinois
  Addieville, 60-61
  Belleville, 66-67, 107
    Square Hotel, 107
  Chicago, 38, 40, 51-53, 64, 67-68, 71, 75, 78-79, 85
    Cicero Field, 38
  Fort Sheridan, 62-63
  O'Fallon, 36
  Peoria, 51-53
  Rockford, 213
    Sundstrand Company, 213, 215
  Scott Army Air Base, 22, 24
  Scott Field, 12-13, 15-16, 19-20, 34-38, 40, 42-46, 48, 50, 52-53, 55, 57, 59, 61, 63-65, 67-69, 84, 89, 94-96, 102, 107-108, 150
    Aerial Weather Recon, 19
    Operations Office, 12, 17-19, 25, 28-29, 31-32, 37, 58-60, 62, 65, 95, 107-108
    Upper Air Recon, 23
    Weather Station Headquarters, 12
  Washburn, 53
  Washington University, 89
  Waukegan, 75
Immelman, 211, 224
Indiana
  Indianapolis, 107
  Terre Haute, 38
Indians, 209
IP (instructor pilot), 31-32
Irvine, General Bill, 168-169

— J —

Jack and Heinz Company, 155
Japan, 159, 183-184
Japanese, 158
Jensen, Pete "Mr. Rough," 88, 90-91, 94-95, 97-98, 105, 117, 132, 139, 212-215, 217-219, 222
Jet age, 203
Johnson, Anson "Johnny," ix
Johnston, Alvin "Tex," xi, 207-210, 215-216, 218, 225
Jones
  _____, 69
  Jack, 146, 149, 151, 155-157, 159, 162, 225

— K —

Kansas, 35, 58, 183-184, 186
  Chanute, 35
  Coffeyville, 197, 215
  Council Grove, xiii
Kansas (cont.)
  Emporia, 188
  Fort Riley, 174-175, 202
  Fort Scott, 35
  Lawrence, 209
  Medicine Lodge, 198
  Salina, 158
    Smoky Hill Air Force Base, 158, 218-219, 221
  Topeka, 231
    Combat Air Museum, 231
  Wichita, 34-35, 44, 46, 70, 126, 139-140, 143-144, 149-150, 158, 162-165, 167-168, 175, 177, 207-209, 212, 215, 220-223, 225, 227, 231
    Beech Field, 35
    Broadview Hotel, 34
    Broadway, 143
    Central Christian Church, 223
    Innes Tea Room, 143
    Knoll Aircraft, 44
    McCleren Hotel, 143
    McConnell Air Force Base, 197, 221, 227
    Municipal Airport, 197, 217, 221
    North Pine Crest Avenue, 223
    Westside Airport, 221
    Yellow Cab, 35
Kelley, Major, 65-68
Kenney
  Fred, 105
  Major, 55-59, 61, 66, 69
Kentucky
  Louisville, 123
Kenyon, Ralph, 94-98, 103, 105, 107-115, 117, 119-123, 125-126
Keohn, Mr. and Mrs., 79, 86
Kerr, Jack, 106
Kleinoder, Second Lieutenant Dutch, 22-25, 29, 33-36, 40-42, 44-46, 48, 50, 52-53, 58, 69-70, 95, 107
Knipe, Commander, James, 119-123, 130-132, 135, 139, 150
Knox, Assistant Secretary, 123
Koenig, Major, 23, 42
Korea, 191, 193-194
Krimsreiter, Hans, 81-83, 86
Kuhn, Lieutenant Colonel Bob, 213-214, 218
Kuhns, Lyle, 218

— L —

Lake Ontario, 92
Lakowitz, Frank, 92, 101-102
Lamson, Bob, 159
Langworthy, Gilmore, 117, 139
Lawson
  Dale, 115
  Mr., 46
Leisy, Cliff J., 177, 183, 196, 198-199, 202
Levier, Tony, ix
Levine, Iky, 73-74, 79
Lew, Jim, 111-112, 115
Lewis
  John, 219
  Slim, 139
License
  Commercial pilot, 80-81, 85-86, 89
  Mechanic, 81, 85-86, 89
Lindbergh, Colonel Charles A., xi, xv, 8, 196-199,

Lindbergh, Charles A. (cont.)
  202-203, 215
Lindbergh run, 51, 53
Linderman-Hoverson (L&H), 79-80
Lippish, Jim "Red," 57-59, 61-62, 64-68
Little Oscar (monkey), 206, 210
Lockheed, 146, 203, 225
Lombard, Dr., 206
Lombard helmet fitting, 210
Louisiana
  Barksdale Air Force Base, 219, 230

— M —

Mabry, Lillian, 109, 127
MacKine, Mack, 90-91, 98, 123-124, 132, 136
Maddon, Carl, 188
Maiersperger, Major W. P., 186
Maine
  Boston, 209
    Harbor, 209
Mariner, Bob, 215
Marriner, Lieutenant, 121, 123, 136
Marquette University, 73
Martin California, 119
Mashman, Joe, 163
Massachusetts
  Gloucester, 209
  Lynn, 203, 208-209
    River Works, 203, 208
  Nahant Point, 209
Mayes, Oliver, 146-147, 151
McCauley propeller, 174
McCoy, Colonel Joe, 194
McDonnell, 137
McDowell, Dr., 3, 48
McEwen, R. H., 157
McNeal, Don, xiii
McVey
  Florene "Flo," 175-176, 192, 194, 197-199, 202-203, 205, 207, 213, 219
  George, 156-157, 176
Mechanized warfare, 66
Merrill, Elliot, 139
Metcalf, Dr. Arthur, 230
Meteorologist, 30, 54-55, 86
Meteorology, 13, 89
MGM, ix
Miner, Jim, 40
Minnesota, 74
MIT, 208
Mission Oil Company, 126
Missouri
  California, 97
  Jefferson City, 35, 67, 107
  Joplin, 107-108
  Kansas City, 34
  Lake of the Ozarks, 108
  Neosho, 67
  Ozarks, 35, 108
  Springfield, 34, 67, 225
  St. Louis, 20, 50, 52-53, 67, 70, 89, 94-96, 99, 102, 107-108, 139
    Lambert Field, 35, 51-52, 95, 207
Model A Ford Coupe, 76
Model T Ford, 8
Mongold, C. H., 157
Montgomery, General, 66
Moss, Dr. Robert, 208

— N —

National Engineering Seminar, 168
Nazi, 66
NBC "Today Show," 229
New Jersey, 203
  Fort Monmouth, 13, 54, 89, 196
    Army Tech School, 89
New York, 50, 77, 84, 87, 108, 125, 129, 168, 196
  Bristol Center, 1, 48-49, 77
  Bristol Valley, 77
  Buffalo, 88, 95-97, 102-103, 105, 107, 138, 227
    Alexander Hamilton Institute, 98
    Consolidated Airport, 94, 105
    Municipal Airport, 88, 102
    University of Buffalo, 98
  Canandaigua, 7
    Academy, 8-9
  Conesus Lake, 49
    McPherson Point, 49
  Elmira, 105
  Finger Lakes, 1, 87, 108
  Hammondsport, xiii, 1
    Glen H. Curtiss Museum of Local History, xiii
  Honeoye Falls, 77, 86, 89, 107
  Kenmore, 89, 105
  LeRoy, 88
    Woodward Airport, 88
  Niagara Falls, 105
  Ontario County, 48
  Rochester, 3, 9, 46-47, 49-50, 87, 89, 103, 163
    Times Union School of Aviation, 9, 89
  Syracuse, 13
New York Life Insurance, 70
North American, xv, 103, 114
North Carolina
  Raleigh, 45
North, John, 110-111, 123-126, 128-131
NRA (National Recovery Administration), 75

— O —

Ohio, 174
  Cincinnati, 213, 218
  Cleveland, 44-46, 78, 107, 155
    Cleveland Hotel, 44, 46
    Cleveland Pneumatics, 44
  Columbus, 44-46, 107
  Dayton, 179, 183, 192, 214
    Wright-Patterson Air Force Base, 183-184, 192
      Air Materiel Command (AMC), 183-186
      Patterson Field, 179
      Wright Field, 138, 150-151, 161, 179, 192, 194, 198-199, 203, 206, 210, 212, 214-215, 217
        Power Plant Laboratory, 180
        Test Center, 209
  Elyria, 78
Oklahoma, 50, 58
  Ardmore, 67
  Bartlesville, 50
  Blackwell, 209
  El Reno, 165
  Fort Sill, 64, 67-68
  Lamont, 165, 167-168
  Lawton, 67-68
  Oklahoma City, 68, 198
  Oklahoma University, 113
  Ponca City, 67

Oklahoma (cont.)
Tulsa, 103, 105, 107-109, 116-117, 122-123, 126-128, 134-135, 139, 141-142
glider club, 139
Oppenheimer, Eric, 223
Oregon, 59
Osler
Bob, 205-206
Scott, 204-205, 207-208, 215
OX-5 Aviation Pioneers Hall of Fame, xvi

— P —
Pacific, 45, 59, 158-159
Packard touring car, 66
Patton, General George, 56, 64-68
Peacock, Jack, 155, 157, 159, 225
Peglo
Colonel John, 16-20, 23, 25-26, 29, 32-33, 37, 42-43, 45, 54-62, 64-66, 69, 96, 98
Fritzie (dachshund), 16-18, 25, 69
Mrs., 69
Pennsylvania
Erie, 78
Lancaster, 220
Pierce
Lloyd, 113, 121-123, 132
Sergeant, 58-61
Pigeons, 26, 62
Pink Cadillac, 47-50
Plane Talk (Boeing), 182
Poland, 103
Polish, 70, 74, 79
Popwell, George, 7
Price, Sergeant, 62

— Q —
Quiet Birdmen, 46, 86, 221

— R —
Radcliff, Warren, 157
Radio system
UHF (ultra high frequency), 54
VHF (variable high frequency), 54
Ragwings, x
Rankin, Major, 148-150
Reed, Jim, 77
Reitherman
Al, 128-129, 131-132, 135, 140-143, 145-146
Blondie, 141
Republican (Council Grove, KS), xiii
Reynolds, Gene, 113
RGE factor (Rate of Gas Expansion), 158
Ridley, Captain Jack, 207, 220
Rime ice, 52
Rivers
Mississippi, 19
Salt Fork (Oklahoma), 165, 167
Robbins, Bob, 144-145, 148, 204-206, 209, 215
Rommel, Field Marshal, 66
Roosevelt, President, 54, 75, 139
Rossie, Norma, 214
Rotelli, Roy, 151, 184-185
Rowley
Billy (nephew), 47
Elton Holcomb "Cactus," x-xi, 90, 94, 106, 139, 212
Geraldine (sister-in-law), 47, 49
Rowley (cont.)
Grace Holcomb (mother), 1-3, 5-6, 8, 13, 43, 48-50, 77-78, 86, 127
Jean (wife), 76-77, 79, 84-89, 94, 97, 105, 107, 109, 117, 127, 140, 142, 167, 206208, 223
See also, Austin, Jean
John (father), 1-6, 8-9, 13, 43, 48, 50, 75, 77, 86, 127
John (son), 80, 86-87, 89, 105, 107-109, 175, 203
Maurice (brother), 1, 3, 9, 47, 49-50
Maurice, Jr. (nephew), 47
Russia, 168
Russian, 199
Rutenburr, A. J., 79

— S —
Schaefer, J. Earl, xi, 132, 140-144, 174-175, 182, 184, 186, 193, 197-198, 202, 208, 212-214
Schairer, George, 221
Scham, Jim, 55
Schimmel, Walt, 227
Schoffield, Captain Niles, 15, 19, 23, 25, 29, 32-34, 37-38, 41-42, 45, 50-51, 53, 55-59, 62, 64, 66, 69, 97
Scott, Clayton, 139
Sellers, Captain, 220
Shaffer, Captain Jack, xi, 213, 218
Shedd, Lieutenant, 23, 34-35
Shepard, Brig. General Horace, 184-185
Showalter
Bernice, 206, 208
N. D., xiii, 139, 141, 144, 205-209, 212-214, 217-218, 220-223, 225
Sivley, Bob, 222, 225
Skelly, Bill, 103, 109-112, 117-119, 121-124, 126, 142
Skelly Oil Company, 103, 109-110, 125
Small, Leonard, 151, 156, 157, 225
Smith, Captain, 20, 23-26, 28-29, 31-32, 42, 66
Sodestrum, Harry, 151
South Dakota
Rapid City, 55, 57-58, 68
Soviets, 224
Spartan Air College, 103, 108, 110, 117
Spartan Aircraft
Company, x, 103, 105, 107, 109-110, 115, 117, 123-130, 132, 135-139, 141-142, 144, 165
Manufacturing, 108
School, 116, 120, 129
Special Advanced Procedural Flying, 38
Squawk sheets, 40, 92, 218
Stanley, Bob, 163, 207-208
Stearman (Boeing), 117, 122, 135, 158
Stein, Jack, 81, 85-86
Stevenson, Lieutenant, 54, 61-62
Stewart, Fred, 113, 128, 132, 135, 139
St. Louis Helicopter Company, 207
Stultz, Captain, 179, 197
Surface to Air Missiles (SAM), 224
Swartout
Bill, 47, 49
Mother, 47, 49
Swiler, Bob, 155, 157

— T —
Talbot, Dave, 151, 157
Tang, Colonel Hun Chan, 172-173, 175, 227
Tank, Al, 71-75, 77-79, 81, 83, 86

Taylor, Dick, xi, 224
Teletype, 20, 51-52
Tennessee
  Knoxville, 45
  Memphis, 45, 137, 142
    Naval Training Base, 137
Test Pilot, ix
Test pilots, ix, 48, 61, 68-69, 73, 87-88, 96-97, 102, 112, 114, 119, 155, 179, 185, 207-209, 216, 218, 223-225, 227
Texas, 69, 188
  Kelly Field, 31, 42
  San Antonio, 95
"The Kansas Story," 229
Theodolite, 56-57, 61, 65
Todd, Trooper Mike, 50
Tolley, Fred, 110-111, 115, 117, 119, 125-126
Tones
  Frank, 1-5, 48, 77
  Lonny, 6
Tornadoes, 67-68, 122
Townsend, Lieutenant Colonel Guy, 179, 182, 216
Trombold, George, 141
Truman, Harry, 139, 142, 151
Turkey, 95, 99
Turkish, 94
  Air Force, 99
  government, 94

— U —

UCLA, 206
United States, 16, 68, 76, 99, 140, 173, 182, 206
  Air Force, 177, 179, 182-188, 191, 194-196, 198-200, 206, 211-215, 217-219, 221-224, 227
  Army, 13, 16, 49, 67, 89, 97, 103, 110, 126, 158, 162-163, 170, 175-176
    Air Corps, x, 13, 25, 27, 36, 43, 51-52, 54-55, 67, 84, 88, 102
      Heavier-than-air, 13, 16, 20-21
      Lighter-than-air, 13, 16, 55, 57, 108
      Signal Corps, 13, 15, 26, 36, 43, 52, 54, 59
    Air Forces, 140, 145, 147, 149, 159, 168, 177
      Southwest Procurement, District, 145, 148
  Bureau of Air Commerce, 70-71, 74-75
  Military Air Transport Command, 108
  National Guard, 38, 51
  National Weather Center, 96
  Navy, ix-x, 16, 90-91, 103, 110-115, 119, 121, 123-124, 128, 130-132, 134-137, 141, 195
    Naval Aircraft Factory, 119
    Ten-turn spin test, 113, 115
  Reserve, 94
  Strategic Air Command, xi
  Strategic Institute, 230
  Weather Bureau, 62, 68
University of Minnesota, 73
Utah
  Wendover, 222

— V —

Vandenberg, General Hoyt S., 196
Vanik, Mel, 205
Vaughn
  Colonel Ralph, 140-141, 145-150
  Dave, 113-115, 119
  Stanley, 88, 101, 106
Vietnam War, 227
Von Zeppelin, Count, 16

— W —

Walkey, Ken, 128-129, 134-136, 142
Warren, Major, 194
Washington, 59, 188
  Boeing Field, 138
  Moses Lake, 203, 205-207, 209-210
    Boeing Test center, 203
  Seattle, xi, 144, 183-184, 187, 193-194, 199, 205-206, 215-216
  Snoqualmie Pass, 46, 206
Washington, D.C., 121-124, 126, 129-131, 134, 168
  Smithsonian Institution, xiii
    National Air and Space Museum, xiii
  Willard Hotel, 122, 131
Weather balloon, 56
  Explorer I, 54-58, 62, 64, 68
    Catenary ring, 56, 60, 62
  Tracker balloon, 57, 59-60
Weather traces, 19-20
Weines, Sergeant "Shorty," 26, 64
Weining, Earl, 103, 110, 112-115, 118-119, 121, 123, 125-126, 139-141, 163, 174
West Point, 52, 174, 193
Wheeler, Mel, 165, 168, 175
White, Ben, 81-83, 86
Whiteside, Lynn, 214-215
Wickham, Jim, 165, 167, 173-174, 199, 208-209
Wilkins, Ted, 119, 124-125, 128, 135
Williams, Bob, 113, 146, 151, 155, 157-158, 163, 176, 196, 198-200, 205
Wilson, Sergeant, 210
Wind-hole, 55-57, 59
Wind tunnel, 81, 83
Wisconsin, 74, 77
  Milwaukee, 46, 62-63, 68, 70-71, 74, 76, 78-81, 89
    Eureka Vacuum Cleaner Company, 71
    Fox Point, 76
    Milwaukee Airways, 81, 89, 110
    Milwaukee Parts Company, 71
    Municiple Airport, 83-84
    Schusters Department Stores, 46, 71, 74, 77
    YMCA, 70, 74-75
Wolfe
  Captain Joe, 220, 222, 225
  General K. B., 179, 184, 193, 196, 198, 203, 212-215, 217, 219, 221
Wolfe Project, 184
Wood, Lysle, 183
World War I, 9, 16, 31, 72, 227
World War II, ix, xv, 54, 66, 68, 106, 111, 130, 147, 163-164, 183, 187, 196, 211, 225, 227

— Y —

Yankee, Paul, 34-35
Yeager, Chuck, x, 207
Young, Ora, xi, 70-71, 75, 83-86

— Z —

Zipp
  Harold W., 139-142, 144, 150-151, 163, 165, 167, 174, 177, 182-184, 193, 197-198, 205-206, 208
  Polly, 167
Zulkie, Ray, 73, 80